W9-BHT-689

Wild Herds

A TIME-LIFE TELEVISION BOOK

Produced in Association with Vineyard Books, Inc.

Editor: Eleanor Graves
Series Editor: Charles Osborne
Senior Consultant: Lucille Ogle
Text Editor: Richard Oulahan
    Associate Text Editor: Bonnie Johnson
    Author: John Neary
    Assistant Editors: Peter Ainslie, Regina Grant Hersey
    Writers: Peter D. Lawrence, James Randall
    Literary Research: Ellen Schachter
    Text Research: Susan R. Costello
    Copy Editors: Robert J. Myer, Greg Weed
Picture Editor: Richard O. Pollard
    Picture Research: Judith Greene
    Permissions: Cecilia Waters
Book Designer and Art Director: Jos. Trautwein
    Art Assistant: Carl Van Brunt
Production Coordinator: Jane L. Quinson

WILD, WILD WORLD OF ANIMALS
TELEVISION PROGRAM
Producers: Jonathan Donald and Lothar Wolff
This Time-Life Television Book is published by Time-Life Films, Inc.
Bruce L. Paisner, *President*
J. Nicoll Durrie, *Business Manager*

THE AUTHOR

JOHN NEARY was a reporter for the Washington Star, a writer for Life for 12 years and the author of two books, *Julian Bond: Black Rebel* and *Whom the Gods Destroy*. He has also written articles on conservation for nature magazines. In 1973, his fascination with wildlife and the out-of-doors led him to settle in the Sangre de Cristo Mountains of New Mexico near an area where herd animals still roam in the wild.

THE CONSULTANTS

WILLIAM G. CONWAY, General Director of the New York Zoological Society, is an internationally known zoologist with a special interest in wildlife conservation. He is on the boards of a number of scientific and conservation organizations, including the U. S. Appeal of the World Wildlife Fund and the Cornell Laboratory of Ornithology. He is a past president of the American Association of Zoological Parks and Aquariums.

DR. JAMES W. WADDICK, Curator of Education of the New York Zoological Society, is a herpetologist specializing in amphibians. He has written for many scientific journals and has participated in expeditions to Mexico, Central America and Ecuador. He is a member of the American Society of Ichthyologists and Herpetologists, a Fellow of the American Association of Zoological Parks and Aquariums and a member of its Public Education Committee.

JAMES G. DOHERTY, as Curator of Mammals for the New York Zoological Society, supervises the mammal collection of approximately 1,000 specimens at the Society's Zoological Park in the Bronx, New York. He is the author of many articles on the natural history, captive breeding and management of mammals. He is a member of the American Association of Mammalogists and a Fellow of the American Association of Zoological Parks and Aquariums.

MARK MacNAMARA is Assistant Curator of Mammals at the New York Zoological Society.

Wild, Wild World of Animals

Wild Herds

Based on the television series
Wild, Wild World of Animals

Published by
TIME-LIFE FILMS

The excerpt from Out of Africa by Isak Dinesen, copyright 1937 by Random House, Inc., renewed in 1965 by Rungstedlundfonden, is reprinted by permission of Random House, Inc. and Putnam & Co., Ltd.

The excerpt from Golden Shadows, Flying Hooves by George B. Schaller, copyright © 1973 by George B. Schaller, is reprinted by permission of Alfred A. Knopf, Inc. and Collins Publishers.

The excerpt from The Whispering Land by Gerald Durrell, copyright © 1961 by Gerald M. Durrell, is reprinted by permission of The Viking Press, Inc.

The excerpt from Theodore Roosevelt's America by Theodore Roosevelt, edited by Farida A. Wiley, copyright © 1955 by The Devin-Adair Co., is reprinted by permission of The Devin-Adair Co., Inc.

The excerpt from The Great Migrations of Animals by Georges Blond, copyright © 1956 by Macmillan Publishing Co., Inc. is reprinted by permission of Macmillan Publishing Co., Inc. and Librairie Arthème Fayard.

The excerpt from Red Deer by Richard Jefferies is reprinted courtesy of Longman Group Limited.

The excerpt from The Mustang by J. Frank Dobie, copyright © 1952 by J. Frank Dobie, is reprinted by permission of Little, Brown and Company.

Contents

Introduction

by John Neary

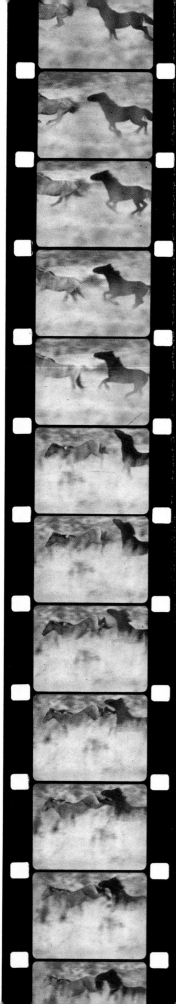

"THE HERD WAS NOT LESS THAN TWENTY MILES IN WIDTH—we never saw the other side—at least sixty miles in length, maybe much longer; two counties of buffaloes! There might have been 100,000, or a million, or a hundred million. I don't know." An early settler's description of a buffalo herd on the Kansas prairie was no tall story. The pioneers of the American West were astounded by the incredible hordes of buffaloes that filled the horizon and sometimes blotted out the sun with the dust they churned up. The awe they created is understandable, for no other four-footed animals that ever lived have gathered together in such numbers. Nor, in all likelihood, will they ever be seen in such numbers again.

Today the great buffalo herds are gone from the plains, and the few captive survivors live on only by the sufferance of man. But spectacular herds, embodying one of the most extraordinary phenomena of animal life, once gathered in many other parts of the world, as seen, for instance, in ancient French cave paintings. In a few places the herds still remain. In the Serengeti Plains of East Africa, for example, upward of 100,000 wildebeests come together each year during the rainy season to begin their long annual migration to greening grazing lands. As the vast horde moves, in one of nature's greatest spectacles, it seems to be a single undulating entity rather than just a huge collection of individual animals that happen to be in the same place at the same time. The herd appears to have an individuality and a life of its own.

What, exactly, is a herd? By a dictionary definiton it is "a number of beasts, especially large animals, clustered together." A herd is certainly that, but so is a troop of baboons, a pride of lions or a pod of whales. Naturalist A. Starker Leopold has defined a herd more precisely: "Any large aggregation, or detached unit, of hoofed animals." Even that definition does not suggest the vital interconnection among the animals that makes a herd a single, disciplined community. For a herd is indeed an entity, surviving because of the individual animals, which in turn owe their own survival to the group. D. R. McCullough, an American ecologist, noted in his study of the tule elk that "the safety of the herd consists of the cohesive mass of animals running in an organized manner. The animals exposed are only those on the outside, and even these are protected by the number of flying hooves and the ebbs and surges within the group. The vast array of movement has a disorienting effect on the observer's vision."

Individual animals get more protection from being part of a herd than just the passive safety of numbers. There can be a greater strength in the whole of which they are just one of the parts, and the whole can do more than just lower the odds that any one of the members may be eaten. Herd animals such as bison and caribou often cooperate to defeat or, at the very least, thwart attacking foes. Water buffaloes in India have been known to charge tigers and literally drive them from the field, sometimes chasing and killing them. Red deer in England divide the task of keeping watch for predators, with the role of sentinel shifting from one hind to another as the herd grazes, so that an undetected approach is impossible.

The herd also magnifies the learning process that takes place between mother

Magdalenian cave painting of bison, Niaux, France

HOOFED HERD MAMMALS
(Ungulates)

ORDER PERISSODACTYLA
(Odd-toed, Hoofed Mammals)

FAMILY EQUIDAE
(Horses, Zebras, Asses)

Common
Zebra

African
Wild Ass

Przewalski's
Horse

This diagram shows hoofed herd mammals grouped in their respective families, each family illustrated by representative animals. All are ungulates with an odd or even number of toes. In the case of the horse and its relatives, which are odd-toed animals and are shown in the upper left-hand corner of this diagram, the axis of the foot passes through the center toe. (The only other odd-toed ungulates are the rhinoceroses and the tapirs, discussed in another volume of Wild, Wild World of Animals.) In even-toed animals, the body weight is distributed between the two middle toes. Even-toed animals are the most diverse and numerous order of ungulates and are found as native species throughout the world except Australia and Antarctica. Families differ widely in their number of species: The Bovidae comprise at least 111 recognized species; the Antilocapridae contain only one species, the pronghorn.

ORDER ARTIODACTYLA
(Even-toed, Hoofed Mammals)

FAMILY TAYASSUIDAE
(Peccaries)

Collared Peccary

FAMILY SUIDAE
(Pigs)

European Wild E

FAMILY CAMELIDAE
(Vicuñas, Guanacos, Camels)

Vicuña

Guanaco

Bactrian Camel

FAMILY CERVIDAE
(Deer)

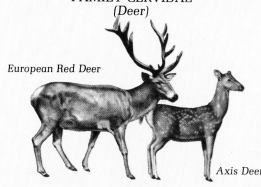

European Red Deer

Axis Deer

FAMILY ANTILOCAPRIDAE
(Pronghorns)

Pronghorn

FAMILY BOVIDAE
(Antelopes, Cattle, Goats, Musk-oxen, Saigas, Sheep)

Saiga

Musk-ox

Cretan
Agrimi Goat

Water
Buffalo

Rocky
Mountain Sheep

Impala
(antelope)

and young, transmitting the cumulative experience of all its members to the young animal as its plays, learns to forage and grows up to challenge the elders for position. Indeed, the herd's cumulative store of knowledge about grazing areas and migration routes is essential to survival. During the exploratory process of evolution the species as a whole derives long-term benefits as a result of herd existence. The plains-dwelling herds afford, through their concentration of females and implacable insistence that males battle each other for the privilege of mating, a far greater opportunity for the furtherance of beneficial traits than would a solitary or familial existence in such an unprotected habitat. The fittest survive and bear young, and the new generations inherit the best characteristics of the herd. Each animal carries the accrued dividends with it even as it continues the process.

The herd creatures described in these pages are ungulates, from the Latin word *ungula*, which means "hoof" or "claw." They are grazers of grass and herbs, browsers of bark and leaves from trees and brush and foragers of nuts, roots, fruits and seeds. Although a few—the wild boar, for example—are not exclusively herbivorous and sometimes feed on smaller creatures, they are much more often the preyed upon. Most of the ungulates have become very swift runners, since flight offers the best chance for survival against numerous predators.

About 60 million years ago, the ancestors of herd animals began emerging in the Eocene epoch as two distinct orders of ungulates—the *Perissodactyla*, or odd-toed, and the *Artiodactyla*, or even-toed. Today the odd-toed ungulates have been reduced to three families—the tapirs, which have three toes, rhinoceroses, which have three, and horses, which have one. Among the 171 species of even-toed animals are antelopes, deer, cattle, sheep, goats, camels, pigs, peccaries and the pronghorn.

It is clear that ungulates, whether odd- or even-toed, evolved toward fewer and fewer toes and developed legs and feet that became lighter over the years. In other words, these were the feet of creatures who were adapting to high-speed locomotion. Today the ungulates are among the fastest creatures on earth.

Ungulates developed still other physical advantages. They have long skulls, which give their noses plenty of room for sensitive olfactory membranes, enabling them to detect an unusually rich range of smells. Their eyes are placed on the sides of their heads, limiting their binocular vision but increasing their peripheral vision. Since most ungulates are herbivores, seeking stationary food, they do not need the binocular vision so necessary to predators. Rather they need a wide-screen view of the ground around them to detect the approach of any enemy. While they rely mainly on their fleetness to escape, they have developed horns and antlers as a means of dominating their own kind and as a defense against predators.

The animals of the herd developed intricate systems of communication that are amazingly subtle and expressive. The rituals by which herd animals mark off their turf and defend it, by which they assume the right to mate and to challenge those that already have it, the biological mechanisms that trigger the onset of the rut, or breeding season, and those that signal the times of the herd's migration are all

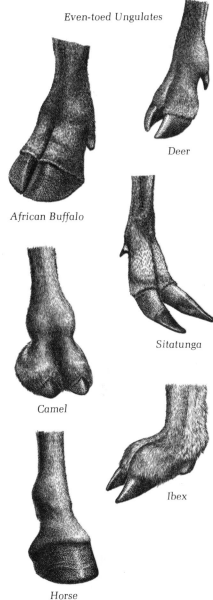

Even-toed Ungulates

Deer

African Buffalo

Sitatunga

Camel

Ibex

Horse
Odd-toed Ungulate

The feet of the hoofed animals shown above are specialized for survival in different environments and against predators. The tiptoed hooves of the deer and the buffalo equip them for speed; the camel's broad, padded foot is suited to travel over sands; the sitatunga's elongated toes help it negotiate swamps; the spongy tissue that makes up the hoof pads of the ibex's foot enables it to cling to cliff faces. Strong but light in weight, the hoof of the horse is engineered for rapid escape.

11

ancient in their origins and all aim toward one end: survival. Territoriality ensures the availability of food; migration serves the same purpose—and is also a rite of passage to more hospitable weather. A breeding season brings forth young at the precise time of the year that is most auspicious for their growth.

These behavioral patterns take on an additional dimension with the feral herds, made up of the descendants of animals once domesticated by man and later freed in some manner, that are able to survive in the wild. In some way, the old, innate herd instincts survived the period of domestication, reasserting themselves when the animals resumed life without man. Domestic cattle, horses, goats and pigs have all been known to take to the wild in feral herds, and all live by the same regimes—similar mating rituals, territoriality, the entire repertoire of herd protocol—that governed the lives of their wild forebears.

Such facts would seem to suggest that the impact of man on these herd animals is only transitory—and, indeed, that is partly true. Many scientists believe that there is no essential difference between the behavior of cattle, horses, pigs or goats, whether they are in the wild or not. For instance, the rancher who turns his stock out of the barn into a large pasture with their fellows creates an environment for them in which they can compete, communicate protect one another and even mate in a manner that would be familiar to their wild or feral counterparts.

In a larger sense, however, man's influence on these creatures is far from negligible. For not one of them, from the presently endangered Rocky Mountain bighorn sheep to the still numerous wildebeest of Africa, is not threatened by man's tendency to enlarge his own realm on earth at the expense of other animals. Survival of the remaining social mammals with which human beings share the earth depends greatly on man's ability to learn that his own life is diminished by the reduction of the herds, both great and small. There is still much to be learned about these creatures and, indirectly, about man as a social animal.

A bison stands alone in the vastness of an American prairie.

Antelopes

The word "antelope" conjures up an image of a swift, graceful animal poised for flight, sniffing the air for the faint scent of danger. Indeed, this image is accurate for most of the 80 species of antelopes: Two well-known species, the impala and gazelle, exemplify athletic grace as do few other animals on earth. But the vision fails to suggest the range of sizes among antelopes—from the largest, the giant eland (weighing up to a ton and standing six feet high at the shoulder), to the smallest, the royal antelope (a mere 10 inches tall and weighing a feathery seven pounds). Nor does the popular image suggest the diverse evolutionary paths followed by the swamp-dwelling sitatunga, equipped with seven-inch-long hooves shaped like bananas, which prevent it from sinking in the mud, and the addax of the Sahara, which lives most of its life without drinking water, getting the moisture it needs from plants, and has disproportionately large feet to prevent it from sinking into the sand.

Derived from a Greek word (antholops) meaning "brightness of eye," the genus name is highly appropriate; "gazelle" comes from an Arabic word, ghazal, also meaning "bright-eyed." Not only luminously beautiful, antelope eyes are also a marvel of efficiency. With their elongated pupils and highly sensitive retinas, the animals, eyes placed on the sides of their heads, are able to keep a sharp watch over most of the territory around them without so much as turning their heads.

The earliest creatures that may have been antelope ancestors, possessing teeth adapted for grazing, first appeared about 23 million years ago in central Asia; 14 million years later, dozens of various species had spread across Europe and Asia. About 100,000 years ago, however, they were driven from the northern regions by cooling weather conditions.

In Africa, where most antelopes live today (some are found in Asia, none in the New World), the habitat can be both dangerous and hospitable. Antelopes are essential to the mammalian food chain—preeminent in the diets of lions, leopards and nearly a dozen other predators, including man. On the other hand, the grasslands of East Africa support, by one estimate, some 30 tons of wild ungulates per square mile. An average of 30 antelopes live on each square mile of the Serengeti Plains, home of 28 antelope species, including 600,000 Thomson's and 40,000 Grant's gazelles, 450,000 wildebeests and 20,000 topi, which share the plains with 170,000 zebras, giraffes, rhinoceroses,

leopards and lions. The antelopes live mostly on grass or leaves, a diet that allows enough forage for each antelope—and enough antelopes to supply their predators.

To survive in the midst of those who prey on them, antelopes need a few advantages, and fortunately nature has equipped them lavishly. They command acute senses of hearing and smell, long legs for rapid flight, exquisitely camouflaged coats and four-chambered ruminant stomachs—a characteristic of many other herd animals that enables them to make the best use of available food.

Almost from the moment of birth the antelope lives on the run. On the open plain a wildebeest calf can stand within three to five minutes after it is born. In another 300 seconds it is loping alongside its mother with the herd, and by the next day it can keep pace with the swiftest adult, whose top speed is 35 miles per hour. The impala, often born in the protection of bush country, can take a relatively leisurely 20 minutes before it first rises to its feet.

In defending their young during this critical period, wildebeest mothers may try to stand between their calves and wild dogs or cheetahs. Others, like gazelles, depend on camouflage and the cover of bushes and trees to safeguard their offspring. Mothers eat the afterbirth and the calf's excrement lest the scent betray the baby's position.

Like many other herd animals, some species of antelopes are territorial, the males staking out property within which they guard harems of females. Oryxes are among the few species—the sable, roan and eland are others—that occasionally use their dagger-sharp horns to ward off marauders. Horns do come into play in rituals of territorial defense, but blood is rarely drawn. These struggles are like jousting matches that merely establish dominance over turf and females without risking lives essential to the future of the herd. Oryxes, for instance, never employ their horns to stab rival males; they just push and shove with their foreheads, striking each other's horns to maintain contact between bouts of shoving.

One observer watched a curious match between two oryx bulls, one of which had somehow lost his horns. Even so, he played out the complete fighting ritual, slashing at his rival's horns and missing by just the amount of space his horns would have occupied if he had had them. Even more notable was the fact that his opponent acted as if the weaponless bull had possessed horns—not pressing his advantage and respecting the hornless male's pretense in an elaborate simulation of combat.

Grant's gazelles

The shy dik-dik (above) is poised for flight in Africa's dry bush country. Little more than a foot tall, the dik-dik has a long fluted proboscis, which it uses to sniff out food and enemies. Its name is derived from the "dik-dik" sound it makes when frightened. Herds of Thomson's gazelles, a two-foot-tall species like the one at right and said to number over half a million animals, migrate westward across the Serengeti Plains each year during the dry season in search of greener pastures.

An Array of Antelopes

If all the world's hoofed animals could be corralled in one place at one time, the antelope would stand out as by far the most numerous and varied member of that large group of mammals. They range in size from the tiny dik-dik (above, left), which is as little as a toy poodle and which lives in monogamous couples or small family groups rather than herds, to the larger Thomson's gazelle (above, right), the smallest herd antelope, to the Goliath of the group, the eland (opposite).

In tune with their extraordinary diversity, antelopes are highly specialized eaters. Each according to its size, behavior and habitat fits into the ecosystem of Africa or Asia. Gerenuks, for example, stand on their hind legs to feed at a level of vegetation utilized by few other animals, while the Thomson's gazelles prefer grazing on grass.

Size, agility and color patterns have also determined the antelopes' methods of protecting themselves. The spindly legged dik-dik is small enough to dart into thick, thorny bushes that hide it and discourage their larger predators. The Thomson's gazelle's great speed and maneuverability enable it to outrace predators and live to breed another day. It is probably the most numerous of the antelopes of Africa.

Among the larger species, the male eland's size generally discourages attacks except from predators working in concert. The smaller female elands are thought to be much more vulnerable to predators.

16

Without its twisted horns and the pale zebralike striping on its sides, the eland would resemble a cow. Indeed the eland's meat is said to taste like that of a cow or steer, which has made it a prime target of some hunters. While on a hunting trip to East Africa Teddy Roosevelt proposed exporting elands to America for butchering. Though the scheme never caught on, the eland is still hunted for its delicious meat in Africa, where its numbers are dwindling.

A Contest for Territory

The males of many antelope species engage in complex, curious rituals designed to gain and hold territory for harems. Some males may mark off territory by defecation or by using specialized glands under their eyes to spread their scent on bushes, trees and grass. Once an antelope, like the spectacular giant sable below, marks off his property, which may encompass as much as 1,000 acres, he must defend it against encroaching males (opposite).

Defense involves a usually bloodless ritual battle in which the combatants butt one another with their formidable horns, which may grow as long as five feet. If the challenged antelope wins, he keeps his turf; if he loses, he becomes a loner or travels in small bachelor groups, in either case waiting for an opportunity to challenge and defeat the dominant male. Driven by instinct, dominant males will defend territory against any challenger.

Locking their long horns, rare giant sables spar in a Luanda game preserve. In their ritual struggle, males will butt, twist and push until one tires and is driven away. Sometimes a battle will end when a horn is broken.

Few animals can boast headgear as spectacular in appearance as the antelopes'. Threat displays and ritualistic battles rather than bloody violence settle most disputes, which erupt over territory or females, though in such displays larger horns often help establish dominance. When a male antelope confronts another of the same species, they usually bang horns, then twist necks in an attempt to throw each other to the ground. The spirals on the horns of certain species are thought by some to be safety devices, evolved to prevent horns from slipping and causing accidental injury.

Uganda kob

Common waterbuck

Eland

Thomson's gazelle

Greater kudu

Gerenuk

Blesbok

21

Escape Artists

The two antelope species pictured on these pages are well equipped to evade predators. One of the most explosive and dazzling of the species is the impala, shown bounding across an East African plain (right) and frozen in mid-leap (below). In order to outmaneuver the large carnivores that prey on them, impalas have evolved heavily muscled thighs and long, slender lower limbs, a combination that produces prodigious jumps up to 35 feet long and 10 feet high. When frightened, members of impala herds leap in several directions, making it difficult for a pursuing enemy to concentrate on singling out one victim.

Wildebeests live on open plains and generally seek the security of large herds to avoid predation. When that fails, however, they take flight, like the wildebeests on the opposite page, galloping head down with sudden swerves and plunges while their whisk-broom tails flick the air. Despite their spindly legs and awkward gait, they can rapidly accelerate to speeds of more than 35 miles an hour.

Horns form a spiky frieze above the heads of oryxes galloping across the dusty East African scrubland. Aside from their horns, which can grow four feet long, East African oryxes have striped faces, black stripes that neatly outline their flanks and black bands around their knees. When there is no water available, oryxes feed on desert vegetation, obtaining much of their moisture from roots that are dug up with their hooves.

25

Out of Africa

by Isak Dinesen

Isak Dinesen, a Danish baroness, lived on a coffee plantation in British East Africa from 1914 to 1931. She later wrote a classic book, Out of Africa, *recording her impressions of the land, the people and the wildlife of the region. In the following excerpt from that book, Dinesen describes the return of Lulu, a bushbuck antelope that she had raised from infancy and that had recently escaped into the bush. After an absence of days, the animal is spotted by Kamante, the household chef.*

One evening Lulu did not come home and we looked out for her in vain for a week. This was a hard blow to us all. A clear note had gone out of the house and it seemed no better than other houses. I thought of the leopards by the river and one evening I talked about them to Kamante.

As usual he waited some time before he answered, to

digest my lack of insight. It was not till a few days later that he approached me upon the matter. "You believe that Lulu is dead, Msabu," he said.

I did not like to say so straight out, but I told him I was wondering why she did not come back.

"Lulu," said Kamante, "is not dead. But she is married."

This was pleasant, surprising, news, and I asked him how he knew of it.

"Oh yes," he said, "she is married. She lives in the forest with her *bwana*,"—her husband, or master. "But she has not forgotten the people; most mornings she is coming back to the house. I lay out crushed maize to her at the back of the kitchen, then just before the sun comes up, she walks round there from the woods and eats it. Her husband is with her, but he is afraid of the people because he has never known them. He stands below the big white tree by the other side of the lawn. But up to the houses he dare not come."

I told Kamante to come and fetch me when he next saw Lulu. A few days later before sunrise he came and called me out.

It was a lovely morning. The last stars withdrew while we were waiting, the sky was clear and serene but the world in which we walked was sombre still, and profoundly silent. The grass was wet; down by the trees where the ground sloped it gleamed with the dew like dim silver. The air of the morning was cold, it had that twinge in it which in Northern countries means that the frost is not far away. However often you make the experience,—I

thought,—it is still impossible to believe, in this coolness and shade, that the heat of the sun and the glare of the sky, in a few hours' time, will be hard to bear. The grey mist lay upon the hills, strangely taking shape from them; it would be bitterly cold on the Buffalo if they were about there now, grazing on the hillside, as in a cloud.

The great vault over our heads was gradually filled with clarity like a glass with wine. Suddenly, gently, the summits of the hill caught the first sunlight and blushed. And slowly, as the earth leaned towards the sun, the grassy slopes at the foot of the mountain turned a delicate gold, and the Masai woods lower down. And now the tops of the tall trees in the forest, on our side of the river, blushed like copper. This was the hour for the flight of the big, purple wood-pigeons which roosted by the other side of the river and came over to feed on the Cape-chestnuts in my forest. They were here only for a short season in the year. The birds came surprisingly fast, like a cavalry attack of the air. For this reason the morning pigeon-shooting on the farm was popular with my friends in Nairobi; to be out by the house in time, just as the sun rose, they used to come out so early that they rounded my drive with the lamps of their cars still lighted.

Standing like this in the limpid shadow, looking up towards the golden heights and the clear sky, you would get the feeling that you were in reality walking along the bottom of the Sea, with the currents running by you, and were gazing up towards the surface of the Ocean.

A bird began to sing, and then I heard, a little way off in the forest, the tinkling of a bell. Yes, it was a joy, Lulu was back, and about in her old places! It came nearer, I could follow her movements by its rhythm; she was walking, stopping, walking on again. A turning round one of the boys' huts brought her upon us. It suddenly became an unusual and amusing thing to see a bushbuck so close to the house. She stood immovable now, she seemed to be prepared for the sight of Kamante, but not for that of me. But she did not make off, she looked at me without fear and without any remembrance of our skirmishes of the past or of her own ingratitude in running away without warning.

Lulu of the woods was a superior, independent being, a change of heart had come upon her, she was in possession. If I had happened to have known a young princess in exile, and while she was still a pretender to the throne, and had met her again in her full queenly estate after she had come into her rights, our meeting would have had the same character. Lulu showed no more meanness of heart than King Louis Philippe did, when he declared that the King of France did not remember the grudges of the Duke of Orleans. She was now the complete Lulu. The spirit of offensive had gone from her; for whom, and why, should she attack? She was standing quietly on her divine rights. She remembered me enough to feel that I was nothing to be afraid of. For a minute she gazed at me; her purple smoky eyes were absolutely without expression and did not wink, and I remembered that the Gods or Goddesses never wink, and felt that I was face to face with the ox-eyed Hera. She lightly nipped a leaf of grass as she passed me, made one pretty little leap, and walked on to the back of the kitchen, where Kamante had spread maize on the ground.

Kamante touched my arm with one finger and then pointed it towards the woods. As I followed the direction, I saw, under a tall Cape-chestnut-tree, a male bushbuck, a small tawny silhouette at the outskirt of the forest, with a fine pair of horns, immovable like a tree-stem. Kamante observed him for some time, and then laughed.

"Look here now," he said, "Lulu has explained to her

husband that there is nothing up by the houses to be afraid of, but all the same he dares not come. Every morning he thinks that to-day he will come all the way, but, when he sees the house and the people, he gets a cold stone in the stomach,"—this is a common thing in the Native world, and often gets in the way of the work on the farm,—"and then he stops by the tree."

For a long time Lulu came to the house in the early mornings. Her clear bell announced that the sun was up on the hills, I used to lie in bed, and wait for it. Sometimes she stayed away for a week or two, and we missed her and began to talk of the people who went to shoot in the hills. But then again my houseboys announced: "Lulu is here," as if it had been the married daughter of the house on a visit. A few times more I also saw the bushbuck's silhouette amongst the trees, but Kamante had been right, and he never collected enough courage to come all the way to the house.

One day, as I came back from Nairobi, Kamante was keeping watch for me outside the kitchen door, and stepped forward, much excited, to tell me that Lulu had been to the farm the same day and had had her Toto,—her baby—with her. Some days after, I myself had the honour to meet her amongst the boys' huts, much on the alert and not to be trifled with, with a very small fawn at her heels, as delicately tardive in his movements as Lulu herself had been when we first knew her. This was just after the long rains, and, during those summer months, Lulu was to be found near the houses, in the afternoon, as well as at daybreak. She would even be round there at midday, keeping in the shadow of the huts.

Lulu's fawn was not afraid of the dogs, and would let them sniff him all over, but he could not get used to the Natives or to me, and if we ever tried to get hold of him, the mother and the child were off.

Lulu herself would never, after her first long absence from the house, come so near to any of us that we could touch her. In other ways she was friendly, she understood that we wanted to look at her fawn, and she would take a piece of sugar-cane from an outstretched hand. She walked up to the open dining-room door, and gazed thoughtfully

into the twilight of the rooms, but she never again crossed the threshold. She had by this time lost her bell, and came and went away in silence.

My houseboys suggested that I should let them catch Lulu's fawn, and keep him as we had once kept Lulu. But I thought it would make a boorish return to Lulu's elegant confidence in us.

It also seemed to me that the free union between my house and the antelope was a rare, honourable thing. Lulu came in from the wild world to show that we were on good terms with it, and she made my house one with the African landscape, so that nobody could tell where the one stopped and the other began. Lulu knew the place of the Giant Forest-Hog's lair and had seen the Rhino copulate. In Africa there is a cuckoo which sings in the middle of the hot days in the midst of the forest, like the sonorous heartbeat of the world, I had never had the luck to see her, neither had anyone that I knew, for nobody could tell me how she looked. But Lulu had perhaps walked on a narrow green deerpath just under the branch on which the cuckoo was sitting. I was then reading a book about the old great Empress of China, and of how after the birth of her son, young Yahanola came on a visit to her old home; she set forth from the Forbidden City in her golden, green-hung palanquin. My house, I thought, was now like the house of the young Empress's father and mother.

The two antelopes, the big and the small, were round by my house all that summer; sometimes there was an interval of a fortnight, or three weeks, between their visits, but at other times we saw them every day. In the beginning of the next rainy season my houseboys told me that Lulu had come back with a new fawn. I did not see the fawn myself, for by this time they did not come up quite close to the house, but later I saw three bushbucks together in the forest.

The league between Lulu and her family and my house lasted for many years. The bushbucks were often in the neighbourhood of the house, they came out of the woods and went back again as if my grounds were a province of the wild country. They came mostly just before sunset, and first moved in amongst the trees like delicate dark silhouettes on the dark green, but when they stepped out to graze on the lawn in the light of the afternoon sun their coats shone like copper. One of them was Lulu, for she came up near to the house, and walked about sedately, pricking her ears when a car arrived, or when we opened a window; and the dogs would know her. She became darker in colour with age. Once I came driving up in front of my house with a friend and found three bushbucks on the terrace there, round the salt that was laid out for my cows.

It was a curious thing that apart from the first big bushbuck, Lulu's bwana, who had stood under the Cape-chestnut with his head up, no male bushbuck was amongst the antelopes that came to my house. It seemed that we had to do with a forest matriarchy.

The hunters and naturalists of the Colony took an interest in my bushbucks, and the Game Warden drove out to the farm to see them, and did see them there. A correspondent wrote about them in the *East African Standard*.

The years in which Lulu and her people came round to my house were the happiest of my life in Africa. For that reason, I came to look upon my acquaintance with the forest antelopes as upon a great boon, and a token of friendship from Africa. All the country was in it, good omens, old covenants, a song:

"Make haste, my beloved and be thou like to a roe or to a young hart upon the mountain of spices."

29

Complementary Cuisines

One of the advantages of antelope diversity is the fact that different species eat bark and leaves from different trees and bushes; they may also take leaves at different heights on the same plants—and among the grazing antelopes, even feed at different levels of the grasses. These preferences complement one another, and allow a great variety of animals to inhabit the same place without overtaxing the environment. Red oat grass, for example, is eaten by both wildebeests and topis, but since each prefers a different stage of growth, neither competes directly with the other.

Certain antelopes, such as the impalas seen below, as well as elands and gazelles, graze during the wet season when young grasses are readily available. At other times, they browse on bark and succulent buds.

Standing on tiptoe, a gerenuk stretches to nibble a morsel from a thornbush. In build the gerenuk is unlike any other antelope. It has an elongated neck and extra-long legs, giving rise to its nickname, the "giraffe-gazelle." Both of these characteristics probably result from the fact that it evolved, like the giraffe, into a specialized tree-feeder. Triangular hooves help brace the gerenuk during its acrobatic search for food.

Kings of the Road

Twice a year shaggy-bearded wildebeests like the ones below perform one of the most awesome parades seen in the wildlife kingdom. At the beginning of the dry season, in early June, vast herds that number up to 100,000 head move in formations (opposite) that stretch for miles from the southern portion of Tanzania's Serengeti Plains to the northwest region, where there is plentiful water and food. In November, the wildebeests turn smartly and, following the gazelles and zebras, rumble back south over hoof-worn trails that from the air appear as highways.

With almost military precision, wildebeests in column wade across a lake (right) in the Serengeti. During their southward migration, the wildebeests attract as camp followers a number of predators and scavengers: lions, jackals, wild dogs, hyenas and vultures, which feed on strayed or dead calves and the older, weaker adults or other animals unable to keep up with the herd.

32

Golden Shadows, Flying Hooves

by George Schaller

While studying lions for his book Golden Shadows, Flying Hooves, *naturalist George Schaller took the opportunity to observe the behavior of other animals on the Serengeti Plains, including the wildebeest. Schaller was fascinated by their migratory movements—in particular, by the urgency and cohesion of the herd on the march.*

The wildebeest is a strangely fashioned antelope, looking as if assembled from the leavings in some evolutionary factory. Its head is heavy and blunt, and it has a shaggy white beard and knobby, curving horns that give it a petulant mien. Its stringy black mane is so sparse that it seems to compensate for this thinness by having several vertical black slashes on its neck, an arrangement comparable to someone simulating a toupee by drawing black lines on his pate. The bulky shoulders give way to spindly hindquarters and a plumed tail that flails about as with a will of its own. Wildebeest alone seem rather woeful, but *en masse* they convey a strange beauty and power. Occasionally I walked among them. Retreating a few hundred feet to let me pass, the animals near me stood silently, their horns shining in the sun. Those farther back continued their incessant grunting, sounding like a chorus of monstrous frogs. Now and then several animals dashed off in apparent panic only to halt and stare back at me. The air was heavy with odor—earth and manure and the scent of trampled grass.

Wildebeest herds on the march are at their most impressive. Trudging along in single file or several abreast, they move in a hunched gait, only to break suddenly into a lope as they pour over hills and funnel down valleys, herd after herd, a living black flood tracing the age-old trails of their predecessors. This immutable urge to stay with the herd, to move in the direction of the others, causes them to press forward regardless of obstacles in their path. One day several lionesses settled by a brushy ravine that had been crossed by several herds during their erratic trek. In the course of a few hours the lionesses captured six wildebeest, yet, after hesitating briefly, each succeeding herd rushed recklessly ahead, deterred neither by the bodies of their compatriots nor the smell of blood and lions. If a river bars their way, they plunge in, disregarding all dangers, and many may drown. One day I watched about a thousand wildebeest gallop in a long line toward the Seronera River, a mindless mass in motion seemingly without reason or purpose. Those in the lead hurled themselves down the embankment, hit the water, and swam to the opposite side where its steepness halted them. The horde swept in behind and soon the water was crowded with thrashing animals, rearing up, climbing over each other, desperate in their attempts to scramble up the slippery sides. Some were pushed under so far that only their noses and bared white incisors were visible as they strained to stay above water. High-pitched bleats reached a frantic crescendo when those in front turned and met the rest still pressing ahead. Finally some gained the far bank and the others surged back past me with frantic, rolling eyes, still racing with implacable urgency except that now they headed in the direction from which they had just come. Seven dark bodies floating silently in the river attested to their violent passing.

Zebras and Horses

Herds of wild horses once thundered across the plains of most of the world's continents, all descendants of a tiny creature scarcely larger than a whippet. Its name was Eohippus, the dawn horse that first appeared in the Eocene epoch about 60 million years ago.

Most of the wild horses of the past, which included at least 20 separate genera, vanished into extinction. Herds that roamed North America had all died out before the white explorers arrived. For all their relative freedom, the herds popularly regarded today as living wild—the mustangs of the West and the spunky ponies of Chincoteague—are not wild but feral (see pages 112–125), descendants of domestic animals that somehow broke loose from man.

The one surviving species of true wild horse is called Przewalski's horse, after Nikolai Przewalski, the Russian explorer who is credited by some for its discovery on the central Asian steppes in 1879. It belongs to the same genus, *Equus*, that includes all domestic horses alive today: the 60 or so recognized breeds of domestic horses, as well as closely related species that are also scientifically classified with the horses. These are the three surviving species of zebras (*Equus burchelli, zebra* and *grevyi*), and one, the quagga, which is extinct; the Asiatic onager or wild ass (*Equus hemionus*), known variously as the kulan, kiang, chigetai and ghorkhar; and the ass (*Equus asinus*). Many species of *Equus* are known to interbreed, but their offspring (for example, the mule, usually a cross between a male ass and a female horse) are generally sterile and are therefore unable to reproduce.

Wild equine herds are vulnerable to the encroachments of man as a hunter and as a farmer. The spread of agriculture, which brought the quagga to extinction, has done nearly the same for the Cape mountain zebra by eliminating it from its native habitat. Wild horses called tarpans existed in Russia as late as the 19th century but died out. Ironically, the last wild horse, *Equus przewalski*, survives today only in zoos. In Russia lives a mare named Orlica that has the distinction of being the only—and probably the last—living member of that species to be captured in the wild. As of 1976, other specimens—numbering 108 males and 146 females—live in 60 collections around the world, most having descended from 11 foals brought to Germany from central Asia in 1901.

Despite the fact that these herds no longer run free, they do offer some clues to the behavior of the Przewalski's and other wild equids (such as the tarpan). Gray-yellow or yellow-brown, their muzzles light, with a short mane stiff in an erect crop without a forelock, Przewalski's stallions still shepherd their mares attentively. A stallion interposes himself between any observer and his herd of a half-dozen or so mares and their foals, guarding them from what he perceives as a potential threat. Then, after sounding a snort of alarm, he drives them all along in single file. He trots around the herd threatening stragglers, his ears laid back, chasing, kicking and biting those who resist his wishes. At feeding time, the entire herd dines together. But if there is any disturbance mares and foals eat first, followed by older colts and fillies, while the stallion stands back, on the alert. During the breeding season, usually in spring, stallions are even more aggressively protective, biting rivals viciously. Mares zealously watch over their foals until they are nearly a year old, but stallions take little interest in their offspring.

Among wild equines, herd behavior may vary. Asses, onagers and zebras live today in nature much as groups of Przewalski's horses do in zoo exhibits. Some herds, however, like the kiang in the highlands of Tibet and the onager of Iran and India—both subspecies of the Asian wild ass—are led by mares. In such herds, adult males stay by themselves except during the breeding season. In one species of zebra, Grévy's, the female and young form loosely knit herds. Males seem to be territorial and live separately. By contrast with Grévy's, plains zebras, the most common and widespread zebras, live in tightly knit social groups. When one of the herd (half-a-dozen mares and their foals led by a stallion) is missing, the others may even launch a search for it.

Because the zebra seems a more primitive form of *Equus* than the horse, scientists believe that horses should be regarded as zebras that have lost their stripes. Just what survival function those striking patterns serve is uncertain. But evidence does exist indicating that the stripes, which are arranged in patterns unique to each species, help members of the herd recognize their fellows by sight. One researcher set up a stuffed zebra and confronted a herd of zebras with it. They approached, ears pricked up in curiosity, inspected its nostrils, neck, withers, flanks and tail, as if it were a real zebra.

A Savanna Society

The amorphous zebra herds—numbering up to 10,000 during the rainy season—roaming the African savanna, like the plains zebras shown above, are made up of much smaller, highly structured and stable zebra groups (right). According to the German ethologist Hans Klingel, who studied East Africa's plains zebras in detail from 1962 to 1965, these groups comprise either all-male bachelor bands or family units, composed of one stallion and up to six mares with foals.

A dominant, or alpha, mare leads the family group's movement, choosing pastures and watering holes, though her authority may be set aside by the stallion, which travels behind or alongside the group. Young mares will often leave the group when they become sexually mature. A young stallion may be forced from the family, as with many other species of herd animals. Or he may leave on his own, between ages one and four, because his mother may have a new foal and his bonds with her are weakened.

Studies in Black and White

Why does the zebra have stripes? One explanation for the vivid markings of some creatures—that the coloration enhances sexual appeal—doesn't apply to zebras because both sexes are striped. Klingel suggests that, since no two individual zebras are marked in precisely the same way, the stripes help zebras recognize one another—essential information in their tightly knit societies. Another theory, utilized by the Navy during World War II when it painted zebra stripes on ships, holds that the stripes act as camouflage, or "disruptive coloration," breaking up the zebra's silhouette and making it less obvious to predators. Indeed, the heat haze rising from the African plains often helps obscure a zebra herd from a distance of three or four hundred yards, and at night zebras are less distinct than solid-colored animals. Some scientists also believe that the stripes probably confuse the distance perception of a predator, giving the fleet-footed zebra, which runs up to 40 miles per hour, a vital split-second head start.

Considered by many to be the most beautiful of zebras, the Grévy's zebra (above) is also the largest, has the narrowest and, therefore, the most stripes of any zebra. Each of the three species of zebras is represented on these pages, and each has its own distinctive pattern of stripes.

The animal known as Burchell's zebra has been extinct since 1910, but numerous subspecies occur, with extremely variable markings. Some of the subspecies, such as the Chapman's (above), have shadow striping between the darker stripes. In other subspecies the background color ranges from white to buff.

A unique gridiron pattern of stripes across the rump, shown in the picture at left, identifies one type of mountain zebra known as Hartmann's. The other type, the Cape mountain zebra, pictured at right, has a distinctive dewlap down the throat.

A zebra mare and her foal stand at rest in a pose typical of plains zebras. Family unit members show great affection and concern for one another and interact constantly, grooming each other, playing or sometimes merely resting their heads on one another's flanks. Special consideration is shown to foals and old or sick members. Even when zebras are fleeing lions, which consume more zebra meat than any other animal in the Serengeti, the pace of the group may be adjusted to accommodate the slowest member.

Zebras and an egret share a meal of grass in the Serengeti. Zebras and other plains animals move in vast, seasonal migrations across the Serengeti in accordance with water and protein availability. The order of the movement—zebras first, wildebeests second and Thomson's gazelles third—is not haphazard. Zebras feed off the coarser, top levels of the grass, leaving the tender stems, leaves and growing shoots for gazelles and the other grazers which follow the zebra.

42

Two stallions engage in neck wrestling, a ritualized form of combat that zebra stallions use to determine dominance. One zebra places his neck over the other's and presses down with all his strength, occasionally even lifting his forelimbs off the ground. Sometimes the losing animal will let his head drop suddenly, pull back and turn the tables, pressing his neck down on his opponent's.

Fighting in earnest, a stallion rears up on his hind legs, kicking and biting at his opponent. Zebra stallions usually get on well with one another. What fighting there is usually occurs when a mare goes into estrus and is approached by interested stallions. Even then zebras, like other herd animals, rarely fight to the death. Injuries are almost always limited to cuts and bruises.

Ancestral Equines

The myriad varieties of horses and donkeys that have been domesticated by man are all descended from two animals: African wild asses, one of which is pictured below, and Eurasian wild horses, represented here by the Przewalski's horse (opposite and right).

African wild asses live in desolate, stony regions where temperatures may rise to 122° F. during the day. Scarcity of water is common in such arid areas, and wild asses can survive without drinking for longer periods than other members of their genus. They live either singly, in bands of up to 10, led by an older female, or occasionally in herds numbering some 100 members.

Named after the Russian explorer who discovered it on the remote border between China and Mongolia, Przewalski's horse is no longer a creature of the wild but survives only in zoos and game parks. Its original habitat on the central Asian steppes was inhospitable—not only dry and hot in summer but bitter cold in winter.

A probable ancestor of the domesticated donkey, an African wild ass lopes across the desert in Ethiopia. Young asses are said to do without any liquid other than their mother's milk for the first six months of their lives, and even as adults they can live through long periods of drought. Attempts to breed these asses in captivity have met with little success.

Camels

Approximately 40 million years ago, in what was then the subtropical forests of North America, there lived a homely, harelike creature whose fossil remains have been identified as *Protylopus*. From this ancestral stem developed a large and highly diversified variety of animals that evolved, about four million years ago, into the animal we know today as the camel. Some of these species migrated north and west across the Bering Strait land bridge into Asia and Africa, settling in regions containing some of the harshest terrain and the most inhospitable climates on the globe.

About two million years ago some of the descendants of those which stayed in North America—by now evolved into the forerunners of today's llamas, alpacas and vicuñas—found their way to the rugged slopes and the thin, chill air of the Andes in South America; meanwhile, their relicts on the northern continent died out.

All of the estimated 15 million camels alive today are tough, rangy animals; but none are tougher than the some 900 truly wild Bactrian, or double-humped, camels that still live in herds in the Gobi. There they survive temperatures that can range from −20° F. to over 100° F. and endure other hardships, including water and food shortages.

The camel has an impressive range of equipment that enables it to survive in the desert where other creatures would perish. Long, thick lashes protect its eyes from the driven sands of the siroccos. When such sandstorms would otherwise make breathing impossible, the camel can partially close its nostrils. Thick pads connect the two toes on each of its feet, giving the animal snowshoelike support on soft sand, traction on steep mountain trails and insulation from hot or cold ground. The camel eats a wide variety of vegetation, including highly alkaline plants that are unpalatable to other animals, ferreting out food with a long snout that is impervious to thorns. It digests its food in a three-chambered stomach.

Heat that would kill a less hardy creature leaves the camel seemingly unfazed. A thick insulating layer of long hair helps protect the camel so well that the animal does not begin to sweat until the temperature of the air around it is high enough to raise its body temperature to 104° F. The camel concentrates most of its body fat in its humps (or single hump, in the case of the dromedary), where the fat acts as an insulator. Body heat is readily released through the skin overnight, when the camel's body temperature falls to 93° F.

When the camel does begin to perspire, the danger of heat exhaustion from dehydration is far less than it would be in man. A person who loses 12 percent of his bodily fluids dies, because most of that liquid is drawn from his blood, which becomes so viscous that it cannot circulate rapidly enough to dissipate body heat through the skin. The camel, however, can lose as much as 27 percent of its body weight in fluids—well over 350 pounds (or 44 gallons) of water in the case of a full-grown 1,400-pound Bactrian—and still survive, because its bloodstream largely retains its water content. To restore its weight, a camel can drink 200 to 300 pounds of water in a brief time.

The last survivors of the wild camels live in herds of one male and several females in the Gobi, wintering in troops of four to six animals, staying close to sources of water. In summer they move out onto the hillsides in search of forage. Bachelor males herd together, and in the breeding season ferocious fights occur when one of them challenges another male for his females. Gestation takes more than a full year. The young camel is born ready for life in the desert. It walks within a few hours of birth and ambles along, keeping up with the herd by its second day.

Of the four camel relatives in South America, the llama has been domesticated as a beast of burden, and the alpaca exists only in herds kept by Andean natives, who harvest the animals' wool. The guanaco, probably similar to the ancestor of the llama, still roams the plateaus in herds. The guanaco is the tallest of wild South American mammals, 43 inches high and weighing 200 pounds. Guanacos live in herds of four to 10 females led by one male, which is zealous in guarding both harem and territory. Despite the perils of life in the Andes, such as the puma, the guanaco is more widespread than the vicuña, which was probably never very common.

Wild bands of vicuñas live high in the crags of the Andes. They grow to be 30 inches high, weigh about 100 pounds and are fawn-colored with a white patch on their throats. Herds of five to 15 females are led by one ferociously territorial male, which defends his turf of some 30 acres from roving male intruders by biting them—or by regurgitating food and spitting it at them, as do other camelids. A baby vicuña, called a *cría*, is able to outdistance a man on foot shortly after birth. Yet, for all their agility and remoteness, vicuñas survive in the wild today mainly because of protective efforts by man after centuries of human exploitation. They have been sought for their fur, which is so fine that the Incas forbade, on pain of death, its use by anyone except royalty.

Dromedary camel

Humps on this Bactrian camel's back stand plump and erect, a sign that the animal is in good health. A camel's humps serve a twofold purpose: They function as storage bins for excess fat that can be metabolized, yielding energy and water. It is also thought that, since the fat, a poor conductor of heat, is limited almost entirely to the humps, the rest of the animal's body is able to cool itself more rapidly.

Muscular nostrils that can be closed against the desert winds prevent sand from blowing into the animal's nose. A ridge extends from each nostril to the camel's cleft upper lip. Their lips are almost prehensile and with them the camel is able to pluck the leaves and twigs it feeds on.

Desertworthy "Ships"

Uniquely adapted through millennia of life in the wild to withstand the harshness of the desert environment, camels have been a vital element in the survival of the desert-dwelling peoples of Asia and Africa since prehistoric times. They transport people and freight; their soft, woolly coats are used to make tents, blankets and clothing; their manure, when dried, serves as fuel; and their milk and meat provide nourishment.

The dromedaries of northern Africa and the Middle East exist today only as domestic animals. They are slender, leggy creatures with dense, short coats that can shield them from temperatures reaching around 130° F. Still to be found living wild in central Asia, their two-humped relatives, the Bactrian camels, have a stockier build with shorter legs and sport long, shaggy coats that insulate them against the region's winter cold, which can drop to −20° F. These coats are shed in the summer, when temperatures soar to over 100° F. Camels traverse their arid terrain with a pacing gait, first advancing both legs on one side of their bodies, then those on the other. This gait produces the side-to-side rocking motion that gave rise to the camels' nickname, "ships of the desert."

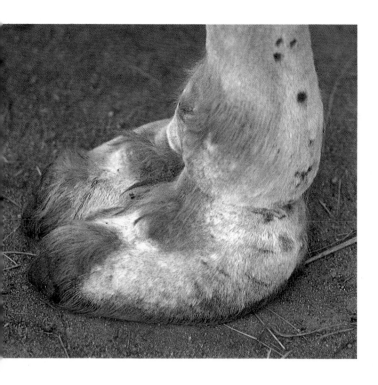

The wide sole of a camel's foot (left) is undivided and thickly padded, giving it a broad base of balance when traveling over the sand. The dromedary above, its eyes protected from wind and sand by thick awnings of interlocking lashes, contorts its face as it chews its cud. When the camel is irritated it spits, a habit that has contributed to the animal's reputation as a difficult, temperamental prima donna.

The African sun beats down on a caravan of nomads and their dromedary camels as they trek across the Libyan Desert. The scarcity of food in the desert necessitates the camel's continual wandering in search of adequate nourishment.

The Camels' Humpless Cousins

The semiarid plateaus of the Andes Mountains are the home of the vicuña and the guanaco, the wild cameloids of the New World. Both species live in herds made up of from four to 15 females, led by a single male. The leader decides when and where the herd will graze (perennial grasses make up most of their diet) and is responsible for the group's safety. When danger appears he emits a shrill whistle, the signal for the females to flee, and positions himself between the intruder and his retreating harem.

Both guanacos and vicuñas live within well-defined territories, the borders of which are marked with urine and dung hills deposited by the male of the herd. The guanaco's coarse coat is of little commercial value, but the demand for the silky wool of the vicuña has resulted in the near extinction of the animal. In an attempt to save these graceful creatures, reserves have been created in Peru and Bolivia where vicuñas like those in the herd above can live undisturbed in the wild.

The vicuña (above) is considered to be among the most graceful of the hoofed animals. The thick, divided cutaneous pads of their small hooves give the vicuñas great agility and sure footing over rocky slopes. Although the vicuña is preyed on by foxes, dogs and pumas, man has, through the ages, been its greatest threat. Today, the vicuña population numbers only about 15,000 animals.

Of South America's two wild cameloids, the guanaco is distinguished from the vicuña by its darker face and much larger size. Guanacos are found in the Andean flatlands from sea level to altitudes up to 14,000 feet. Clearly visible in this open terrain, the guanacos have depended on their alertness and speed, which can exceed 35 miles per hour, for survival.

The Whispering Land

by Gerald Durrell

An inveterate animal collector and incomparable raconteur of animal lore and fact, Gerald Durrell has traveled all over the world in search of rare animals to add to his private zoo, which he maintains on the Isle of Jersey in the Channel Islands. The Whispering Land, *excerpted here, is Durrell's account of his exploration of the tropical forests and shores of the Patagonian region of Argentina. In the following passage, Durrell describes how he discovered,, on the Peninsula Valdes, a herd of wild guanacos that were surprisingly unlike the sorry zoo specimen that represented his only previous encounter with the creature.*

Not only had the landscape changed in colouring and mood but it had suddenly become alive. We were driving down the red earth road, liberally sprinkled with backbreaking potholes, when suddenly I caught a flash of movement in the undergrowth at the side of the road. Tearing my eyes away from the potholes I glanced to the right, and immediately trod on the brakes so fiercely that there were frenzied protests from all the female members of the party. But I simply pointed, and they became silent.

To one side of the road, standing knee-deep in the yellow bushes, stood a herd of six guanacos, watching us with an air of intelligent interest. Now guanacos are wild relatives of the llama, and I had been expecting to see something that was the same rather stocky shape as the llama, with a dirty brown coat. At least, I remembered that the one I had seen in a Zoo many years before looked like that. But either my memory had played me false or else it had been a singularly depressed specimen I had seen. It had certainly left me totally unprepared for the magnificent sight these wild guanacos made.

What I took to be the male of the herd was standing a little in front of the others and about thirty feet away from us. He had long, slender racehorse legs, a streamlined

body and a long graceful neck reminiscent of a giraffe's. His face was much longer and more slender than a llama's, but wearing the same supercilious expression. His eyes were dark and enormous. His small neat ears twitched to and fro as he put up his chin and examined us as if through a pair of imaginary lorgnettes. Behind him, in a tight and timid bunch, stood his three wives and two babies, each about the size of a terrier, and they had such a look of wide-eyed innocence that it evoked strange anthropomorphic gurgles and gasps from the feminine members of the expedition. Instead of the dingy brown I had expected these animals almost glowed. The neck and legs were a bright yellowish colour, the colour of sunshine on sand, while their bodies were covered with a thick fleece of the richest biscuit brown. Thinking that we might not get such a chance again I determined to get out of the Land-Rover and film them. Grabbing the camera I opened the door very slowly and gently. The male guanaco put both ears forward and examined my manoeuvre with manifest suspicion. Slowly I closed the door of the Land-Rover and then started to lift the camera. But this was enough. They did not mind my getting out of the vehicle, but when I started to lift a black object—looking suspiciously like a gun—to my shoulder this was more than they could stand. The male uttered a snort, wheeled about, and galloped off, herding his females and babies in front of him. The babies were inclined to think this was rather a lark, and started gambolling in circles, until their father called them to order with a few well-directed kicks. When they got some little distance away they slowed down from their first gallop into a sedate, stiff-legged canter. They looked, with their russet and yellow coats, like some strange gingerbread animals, mounted on rockers, tipping and tilting their way through the golden scrub.

As we drove on across the peninsula we saw many more groups of guanacos, generally in bunches of three or four, but once we saw a group of them standing on a hill, outlined against a blue sky, and I counted eight individuals in the herd. I noticed that the herds were commoner towards the centre of the peninsula, and became considerably less common as you drove towards the coast. But wherever you saw them they were cautious and nervous beasts, ready to canter off at the faintest hint of anything unusual, for they are persecuted by the local sheep-farmers, and have learnt from bitter experience that discretion is the better part of valour.

Tamed but Rugged

The alpaca (above) and the llama (right) have been domesticated by the Indians of the Andes since the third century B.C., the alpaca for its fine wool, the llama as a beast of burden. Male llamas more than three years old are deemed the best pack animals. Females are kept for breeding purposes and for their wool, which is shorn regularly. Carrying 100-pound loads on their backs, flocks of llamas can cover between 15 and 20 miles a day over the Andean *altiplano*, or high plateau.

Alpacas, with coats that are even longer, thicker and finer than those of llamas, are one of South America's most important wool producers, although they can be shorn only three or four times in a lifetime—once every two years. They have recently been bred with their wild relatives, the vicuñas, and their offspring, the paco-vicuña, bears valuable fleece that combines the quantity yielded by the alpaca with the quality of the vicuña's wool.

56

Pigs and Peccaries

The members of the family Suidae, to give pigs their formal biological name, are the oldest and the most primitive of all the living even-toed ungulates. Pigs have survived so many millions of years because they have adapted to a variety of habitats and an almost omnivorous diet. They are incredibly tough and wily—and, when pushed into a corner or challenged, fearsome.

Pigs have been domesticated since some daring farmer first summoned the nerve either to trap a wild boar and a sow or to raise a piglet from birth. Some scientists believe that this may have happened sometime in the Neolithic era, about 6,500 years ago. Domesticity in a pig, however, is still only a thin veneer over its quintessential wildness. Pigs readily revert to the ways of their wild forebears if they manage to get loose from their pens. In 1912 an enterprising sportsman named George Gordon Moore imported into the United States 14 wild boars from Germany, intending to keep them fenced in on a 1,500-acre hunting preserve in the Appalachian Mountains in North Carolina. A special enclosure constructed of chestnut rails awaited the boars, but even this pen was unavailing against their enormous strength. All the boars broke loose, and the escapees mated with unpenned domestic sows belonging to local farmers and founded a hardy population of some 1,200 hybrid wild boars that now roam the thickly forested ridges and hollows of the mountains just as their ancestors did in Europe.

The European wild boar is a formidable creature indeed, growing to nearly a yard in height and weighing up to 330 pounds, a powerhouse of muscle preceded wherever it goes by razor-sharp tusks some six inches long. The wild boar is an excellent swimmer and will cross rivers and even lakes to escape an enemy or to search for food. It has been hunted out of existence in the British Isles but can still be found in parts of Europe, Asia and North Africa.

The boar will eat just about anything, and some use their tusks in digging, following their keen sense of smell for such buried delicacies as edible roots and shoots, as well as worms and insect larvae. Like all pigs, the wild boar is gregarious—as many as 15 of them bed down in a wriggly mass to snuggle together on cold winter nights—although adult males usually remain solitary except during the breeding season, a period that lasts for three months. Expectant sows build nests, lining them with leaves.

Pigs are prolific: Columbus brought just eight hogs with him to the West Indies in 1493, and 13 years later their offspring were so numerous and so dangerous—herds of them were killing cattle—that islanders mounted organized hunts against them. Hernando de Soto, too, provisioned his expedition to the New World with pork on the hoof, bringing 13 hogs with him to Florida in 1539. Three years later his herd numbered 700 and is now thought to live on as the razorbacks that roam the woods of the South.

In the tropical forests of Africa and Madagascar live the large bush pigs, weighing up to 200 pounds and standing more than 30 inches high at the shoulder. They too are omnivorous but prefer vegetation. They forage at night and sleep through the hot part of the day. The Cape bush pig is fond of going into tilled fields for crops, burrowing under fences or snapping barbed wire to do so.

The largest living pig is the giant forest hog of central Africa, a relatively shy creature, standing a yard high, weighing 300 to 500 pounds and having tusks nine inches long. Forest hogs tunnel runways through the undergrowth from large lairs occupied by families for generations.

The peccary, a member of the family Tayassuidae, is the New World counterpart of the Old World wild swine, somewhat similar in appearance but distinguished by a scent gland on its back that it uses to mark territory and to establish friendly relations with other peccaries through mutual recognition. The family includes two long-known species and one just recently rediscovered. Collared peccaries and white-lipped peccaries live in tropical areas of North, Central and South America. Just a few years ago Dr. Ralph M. Wetzel, a biologist, discovered a species of peccary, thought to have been extinct since the Ice Age, surviving in the wilds of Paraguay. Larger than either the collared or the white-lipped, this third species, dubbed "long-nosed" by Wetzel, is believed to have descended from peccarylike ancestors that migrated to the western hemisphere from Asia over the land bridge that spanned the Bering Strait during dry periods in the age of glaciers. Peccaries have been tamed but never thoroughly enough to be completely trustworthy. Two of them were reported to have served loyally as mascots for a college—until, surfeited with academic life, their innate wildness asserted itself and they gored the institution's president.

Wild boars

Resourceful and Hardy

Wild pigs are among the most adaptable of all mammals. Equally at home in European forests, African plains and Asian scrub, they all require a local body of water (a neighborhood puddle will do). In addition to drinking water, this furnishes the necessary bathing and mud wallowing that keeps them cool and protects them from parasites. Lacking few natural predators, such as wolves, which have been killed off by man, wild pigs, such as the European wild boar (below), multiply prolifically if unchecked by man. In densely populated areas they can become serious pests, causing considerable damage to crops and fields. Regulated hunting keeps their numbers under control.

Destructive though they can be when unchecked, wild boars also play a positive role in their environment. In their search for food, wild boars root up and loosen forest soil, burying seeds in the process and thus contributing to the regeneration of trees and other vegetation.

Wild boar youngsters (above) huddle together, seeking the close body contact all pigs enjoy when resting. They still sport the soft, striped coat that boars have until about three months of age, when it is replaced by a dark, dense fur overlaid by bristles, like that of the adult at right.

The creature at right is one of nearly 200,000 collared peccaries (also known as javelinas) that still roam the New World from Texas to South America in bands of five to 15 males and females of all ages. The range of the collared peccary overlaps that of the larger white-lipped peccary, which stretches as far south as Argentina. Through extensive hunting for its meat and hide, the collared peccary has decreased in numbers in recent years. Its serious enemies, other than man, include the jaguar, bobcat and cougar.

Head held high, back arched and its white mane conspicuous, this bush pig assumes a favorite pose for impressing other bush pigs. Found from the region south of the Sahara to the island of Madagascar, the numerous subspecies of bush pigs live on roots, grass and fallen fruits. Inhabiting watery, well-vegetated areas such as reed beds, they sleep during the heat of the day and forage at night.

Bison and Buffaloes

The bison, buffalo, gaur and yak are among the largest herd animals. But though their size and herding habits have protected them from most predators, these giants have suffered badly at the hands of man, whose tendency to exploit some and evict others from their ranges has driven them all close to extinction.

Few wild animals have undergone a more dramatic and shocking collision with man than the bison—and lived to make a comeback. Once found from Siberia to the European shores of the Atlantic, the bison dispersed across the Bering Strait land bridge to the New World about a half-million years ago. Long before the first white men arrived and misnamed them buffaloes, these shaggy, formidable creatures (which can grow as tall as six feet at the shoulder and weigh a ton and a half) flourished in herds that may have totaled 50 million head. In the late 19th century Americans killed the bison in wholesale numbers for its hide and meat. By 1889, conservationist Dr. William T. Hornaday, later to become the leader of the American Bison Society, estimated that the number of bison left in the United States totaled only 541. Through the conservation efforts of Dr. Hornaday's Society (founded at New York's Bronx Zoo in 1905) and other concerned groups, there are about 36,000 buffalo alive today.

The bison is highly curious, especially about newborn calves and disabled adults. During the great 19th-century slaughter, this curiosity about fallen fellows may have hastened the carnage. Attracted by the smell of a dead bison, a herd companion may excitedly sniff the corpse and push at it with its head, apparently trying to make the stricken animal rise. Hunters could easily pick off these lingering survivors at a kill.

In the Old World, at the time the International Association for the Preservation of the European Bison was formed in 1923, there were just 56 European bison (also called wisents) alive—all in zoos. The last wild one had been killed during World War I. Today, however, there are 2,000 wisents alive in Europe, including two herds in the forests of Poland and the Soviet Union and 1,500 others in zoos.

Nearly as large as the bison, the gaur of southern Asia faces the threat of extinction largely as a result of the intrusion of man's settlements into its habitat—jungle clearings and shady hill forests. In Malaysia, for example, almost all gaur—known there as the seladang—have disappeared. A representative herd may contain about 70 members traveling in groups of from 10 to 20 animals.

The two species of true buffaloes—the African and the Asian—are also dwindling in numbers. Named for its habit of often standing muzzle-deep in water, the water buffalo of Asia is a favorite quarry of hunters, who prize it for its flesh and for the trophy provided by its majestic horns, which can be six feet in length. Though it has been domesticated as a draft animal, the Asiatic buffalo, also standing up to six feet at the shoulder, can be extremely dangerous. In areas where it still survives in the wild (living in small groups of 10 or 20 individuals) the animal moves with the unconcern of a creature that has been known to attack a tiger to drive it away from the herd.

African Cape buffaloes were nearly exterminated by a plague that originated among domestic cattle in the last decade of the 19th century. But some still live in herds along the east coast. Like their Asian kin, these buffaloes are seldom prey to any attackers except man. However, buffaloes have reportedly killed more big-game hunters than have any other animals.

Most inaccessible of the wild cattle are the yaks, which roam the remote steppes and mountains of northern Tibet. There they maintain a population of about 8,000, seriously threatened by Tibetan wolves and nomadic tribesmen, who kill the animals for their meat, hides and tails (the latter are believed to possess magical properties).

As grazers, all of these animals share, with antelopes, camels and deer, one of the most curious survival mechanisms in the animal kingdom—a multichambered, ruminant stomach. Cud chewing is the visible sign of this remarkable adaptation—the mastication of food that has already been partly digested. The process serves two purposes. First, rumination allows almost constant digestion of large amounts of food, eaten when the animal feels secure, or of small snacks nibbled on the run. Digestion can proceed when the animal is on the alert and not feeding or when it is safely resting (a good time for cud chewing). Second, the ruminant stomach allows the highly efficient digestion of low-grade food in great quantities through the action of bacteria that reside in the largest chamber of the stomach, the rumen. In ruminants, these bacteria break down the tough, relatively indigestible cellulose that constitutes the plants' cell walls. Equipped with a small and simple stomach, a horse eats grass and brush but is not as efficient in digesting it as is a ruminant. A human being, assuming he could choke the food down, would get almost no nourishment at all.

African buffalo

Prairie Giants

The grasslands of North America, from the Mississippi to the Rocky Mountains, were once the domain of the prairie bison (below). Each year enormous herds of these beasts, the largest land mammals in North America, slowly made their way across the prairies, following established routes in search of grasses and water. Today the bison survives only on private ranches and in national parks. The herd at right lives in Custer State Park in South Dakota.

Although their range is restricted, today's bison live much as their ancestors did. The herd can vary in size from a family unit of two adults and their offspring to thousands of animals. During the mating season, which begins in July, battles erupt between mature bulls for females. The victor asserts his supremacy by mating with the cows. Calves are born after a nine-month gestation period and may stay with their mothers for three years.

THEODORE ROOSEVELT'S AMERICA

by Theodore Roosevelt

Theodore Roosevelt donned his hunting outfit, drew his rifle and posed in front of a painted backdrop for this portrait taken in 1884.

By the time Theodore Roosevelt was elected President in 1901, concern over the fate of the American bison had already been aroused among conservationists.

Roosevelt's brief history of the bison, which he called the buffalo in the old frontiersman's usage, reflects his ambivalent attitude toward conservation. A hunter, he was excited by the thrill of a buffalo chase; a progressive, he respected the "manifest destiny" of America to spread westward; a conservationist and animal lover, he bemoaned the "necessity" of the slaughter of the noble bison.

When we became a nation in 1776, the buffaloes, the first animals to vanish when the wilderness was settled, roved to the crests of the mountains which mark the western boundaries of Pennsylvania, Virginia, and the Carolinas. They were plentiful in what are now the States of Ohio, Kentucky, and Tennessee. But by the beginning of the present century they had been driven beyond the Mississippi, and for the next eighty years they formed one of the most distinctive and characteristic features of existence on the great plains. Their numbers were countless—incredible. In vast herds of hundreds of thousands of individuals, they roamed from the Saskatchewan to the Rio Grande and westward to the Rocky Mountains. They furnished all the means of livelihood to the tribes of Horse Indians, and to the curious population of French Metis, or Half-breeds, on the Red River, as well as to those dauntless and archetypical wanderers, the white hunters and trappers. Their numbers slowly diminished, but the decrease was very gradual until after the Civil War. They were not destroyed by the settlers, but by the railways and the skin hunters.

After the ending of the Civil War, the work of constructing the transcontinental railway lines was pushed forward with the utmost vigor. These supplied cheap and indispensable, but hitherto wholly lacking means of transportation to the hunters; and at the same time the demand for buffalo robes and hides became very great, while the enormous numbers of the beasts, and the comparative ease with which they were slaughtered, attracted throngs of adventurers. The result was such a slaughter of big game as the world had never before seen; never before were so many large animals of one species destroyed in so short a time. Several million buffaloes were slain. In fifteen years from the time the destruction fairly began the great herds were exterminated. In all probability there are not now, all told, five hundred head of wild buffaloes on the American continent; and no herd of a hundred individuals has been in existence since 1884.

The first great break followed the building of the Union Pacific Railway. All the buffaloes of the middle region were then destroyed, and the others were split into two vast sets of herds, the northern and the southern. The latter were destroyed first, about 1878; the former not until 1883. My own chief experience with buffaloes was obtained in the latter year, among small bands and scattered individuals, near my ranch on the Little Missouri. . . .

The extermination of the buffalo has been a veritable tragedy of the animal world. Other races of animals have been destroyed within historic times, but these have been species of small size, local distribution, and limited numbers, usually found in some particular island or group of islands; while the huge buffalo, in countless myriads, ranged over the greater part of a continent. Its nearest relative, the Old World aurochs, formerly found all through the forests of Europe, is almost as near the verge of extinction, but with the latter the process has been slow and has extended over a period of a thousand years instead of being compressed into a dozen. The destruction of the various species of South American game is much more local and is proceeding at a much slower rate. It may truthfully be said that the sudden and complete extermination of the vast herds of the buffalo is without parallel in historic time.

No sight is more common on the plains than that of a bleached buffalo skull and their countless numbers attest the abundance of the animal at a time not so very long past. On those portions where the herds made their last stand, the carcasses, dried in the clear, high air, or the mouldering skeletons, abound. Last year, in crossing the country around the heads of the Big Sandy, O'Fallon Creek, Little Beaver, and Box Alder, these skeletons or dried carcasses were in sight from every hillock, often lying over the ground so thickly that several score could be seen at once. . . .

Thus, though gone, the traces of the buffalo are still thick over the land. Their dried dung is found everywhere, and is in many places the only fuel afforded by the plains, their skulls, which last longer than any other part of the animal, are among the most familiar of objects to the plainsman; their bones in many districts so plentiful that it has become a regular industry, followed by hundreds of men (christened "bone hunters" by the frontiersmen), to go out with wagons and collect them in great numbers for the sake of the phosphates they yield; and the Bad Lands, plateaus, and prairies alike, are cut up in all directions by the deep ruts which were formerly buffalo trails.

These buffalo trails were made by the herds travelling strung out in single file, and invariably taking the same route each time they passed over the same piece of ground. As a consequence, many of the ruts are worn so deeply into the ground that a horseman riding along one strikes his stirrups on the earth. In moving through very broken country they are often good guides; for though the buffalo can go easily over the roughest places, they prefer to travel where it is smooth, and have a remarkable knack at finding out the best passage down a steep ravine, over a broken cliff, or along a divide. In a pass, or, as it is called in the West, "draw," between two feeding grounds, through which the buffalo were fond of going, fifteen or twenty deep trails may be seen; and often, where the great beasts travelled in parallel files, two ruts will run side by side over the prairie for a mile's length. These old trails are fre-

In 1883 a dashing Theodore Roosevelt was photographed with his horse on his ranch in the North Dakota Badlands.

quently used by the cattle herds at the present time, or are even turned into pony paths by the ranchman. For many long years after the buffalo die out from a place, their white skulls and well-worn roads remain as melancholy monuments of their former existence.

The rapid and complete extermination of the buffalo affords an excellent instance of how a race that has thriven and multiplied for ages under conditions of life to which it has slowly fitted itself by a process of natural selection continued for countless generations, may succumb at once when these surrounding conditions are varied by the introduction óf one or more new elements, immediately becoming the chief forces with which it has to contend in the struggle for life. The white man entered upon the scene, where its phenomenal gregariousness —surpassed by no other four-footed beast, and only equalled, if equalled at all, by one or two kinds of South African antelope—its massive bulk, and unwieldy strength. The fact that it was a plains and not a forest or mountain animal was at that time also greatly in its favor. Its toughness and hardy endurance fitted it to contend with purely natural forces: to resist cold and the winter blasts, or the heat of a thirsty summer, to wander away to new pastures when the feed on the old was exhausted, to plunge over the broken ground, and to plough its way through snowdrifts or quagmires. But one beast of prey existed sufficiently powerful to conquer it when full grown and in health; and this, the grizzly bear, could only be considered an occasional foe. The Indians were its most dangerous enemies, but they were without horses, and their weapons, bows and arrows, were only available at close range; so that a slight degree of speed enabled a buffalo to get out of the way of his human foes when discovered, and on the open plains a moderate development of the senses was sufficient to warn them of the approach of the latter before they had come up to the very close distance required for their primitive weapons to take effect. Thus the strength, size, and gregarious habits of the brute were sufficient for a protection against most foes; and a slight degree of speed and moderate development of the senses served adequate guards against the grizzlies and bow-bearing foot Indians. Concealment, and the habit of seeking lonely and remote places for a dwelling, would have been of no service.

But the introduction of the horse, and shortly afterward the incoming of the white hunters carrying longe-range rifles, changed all this. . . .

The incoming of the cattlemen was another cause of the completeness of their destruction. Wherever there is good feed for a buffalo, there is good feed for a steer or cow; and so the latter have penetrated into all the pastures of the former; and of course the cowboys follow. A cowboy is not able to kill a deer or antelope unless in exceptional cases, for they are too fleet, too shy, or keep themselves too well hidden. But a buffalo neither tries nor is able to do much in the way of hiding itself; its senses are too dull to give it warning in time; and it is not so swift as a horse, so that a cowboy, riding round in the place where cattle, and therefore buffalo, are likely to be, is pretty sure to see any of the latter that may be about, and then can easily approach near enough to be able to overtake them when they begin running. The size and value of the animal made the chase after it very keen. Hunters will follow the trail of a band for days, when they would not follow that of a deer or antelope for a half-hour. . . .

While the slaughter of the buffalo has been in places needless and brutal, and while it is greatly to be regretted that the species is likely to become extinct, and while, moreover, from a purely selfish standpoint, many, including myself, would rather see it continue to exist as the chief feature in the unchanged life of the Western wilderness; yet, on the other hand, it must be remembered that its continued existence in any numbers was absolutely incompatible with anything but a very sparse settlement of the country; and that its destruction was the condition precedent upon the advance of white civilization in the West, and was a positive boon to the more thrifty and industrious frontiersmen. Where the buffalo were plenty, they ate up all the grass that could have supported cattle. The country over which the huge herds grazed during the last year or two of their existence was cropped bare, and the grass did not grow to its normal height and become able to support cattle for in some cases two, in others three, seasons. Every buffalo needed as much food as an ox or cow; and if the former abounded, the latter perforce would have to be scarce. . . .

From the standpoint of humanity at large, the extermination of the buffalo has been a blessing. The many have been benefited by it; and I suppose the comparatively few of us who would have preferred the continuance of the old order of things, merely for the sake of our own selfish enjoyment, have no right to complain. . . .

Indians swirl around a mass of bison in this oil painting, entitled "Buffalo Chase: A Surround by the Hidatsa," done in 1832 by George Catlin, the eminent American artist best known for his portraits of the Indians of North America. The Hidatsa inhabited the upper Missouri River area of North Dakota, where they held organized annual buffalo hunts.

THE GREAT MIGRATIONS OF ANIMALS

by Georges Blond

In the days of their mass migrations, bison usually followed their leaders blindly, heedless of obstacles and pitfalls, and often unintentionally committed mass suicide. Georges Blond, author of the The Great Migrations of Animals, *had heard many stories of such disasters and in the excerpt below recounts the fate of a herd, overtaken on the march by a blizzard in 1858.*

For two days such a snowstorm had been raging, characterized by violent wind, a uniformly gray sky and only a few yards of visibility. The wolves followed the herd at a distance of less than thirty feet, guided by smell. And what could the front line of buffalo see? Only a whirling mass of snow. How, then, were they guided? They did not choose their destination; they submitted to it. Amid our ignorance of the migrant buffalo's sense of direction, we can lay hold of one definite fact. Every time a snowstorm

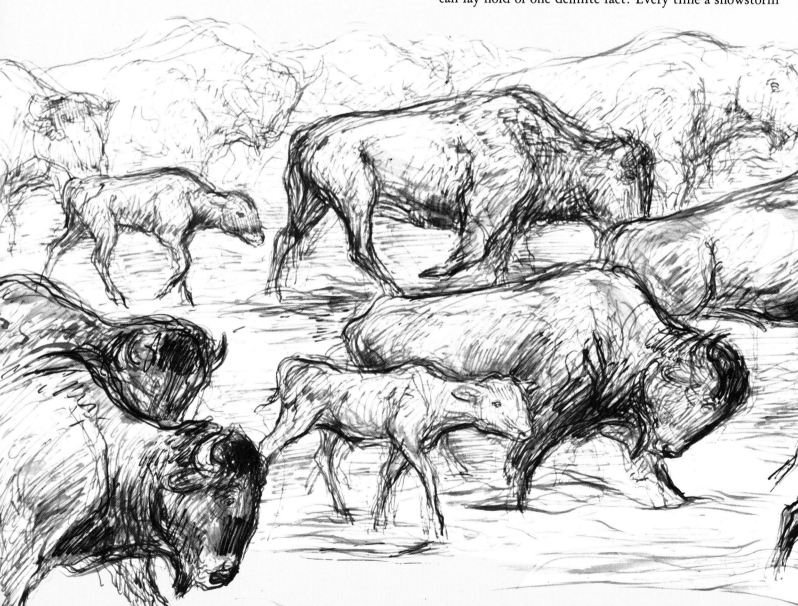

transformed the landscape with its white curtain, the buffalo spontaneously broke into a trot, pushing against the wind. They did not pause to wonder what they should do. The story I am about to tell is of a hundred thousand buffalo in 1858, and the same thing has happened to their kind many times in the course of the ages.

The buffalo trotted at first, and then broke into a sort of single-foot. Although this gait made the whole herd appear to limp, swaying from side to side, it rested them from the effort of breasting the snow. Eventually they slowed down to a walk. The snow stood higher and higher on the ground and swept more and more powerfully out of the sky. Moreover, night was about to fall. None of the buffalo knew the old stories of entire herds that had been found frozen to death in the snow, but many of them had

lived through hard winters and had some inkling of the fact that the approaching night would be anything but gay. It was indeed a night of horror.

The instinct of self-preservation forbade the herd to stop, much less to lie down. All night long they moved on, often fending the snow with their muscular chests. Needless to say, they moved very, very slowly, huddled all together. Behind, the wolves were so close that sometimes their muzzles touched the hind legs of the buffalo. In spite of gnawing hunger they had not the slightest urge to bite. The terrible and imperious snow would not allow them to waste a single motion outside the struggle to keep alive, to keep moving, for fear of being suffocated and buried in the darkness.

Through sheer obstinacy, the buffalo did not perish that

night. When day broke the herd was still together, pushing against the snow. The sun must have been for some time above the horizon, but it was impossible to say exactly where, and the daylight was no brighter than dawn. Toward the middle of the day the storm diminished in fury and the snow stood less high on the ground, so that the buffalo could trot again. There was still snow in the air but there was a visibility of at least thirty feet, such as they had not enjoyed since early the preceding day. The hoofs of the leaders scattered the snow almost joyfully, and the rest of the herd quickened their pace to keep up.

What gave the front-line buffalo a warning? Did they perceive a different coloring in the wall of snow or a variation in its furious movements? At any rate, they slowed down, all together, or attempted to do so. Their legs stiffened, and there was something like a rip-tide in the snow tossed up from the ground. If there had been only one, or at most a few lines of buffalo, they might have cut down their speed before the precipice. But the pressure of the thousands of trotting locomotives behind them did not permit them to stop within the thirty feet of visibility ahead. Like a carpet, the whole herd rolled over the cliff. Nothing could avert the total catastrophe.

Perhaps some of the buffalo heard the noise produced by their predecessors, as they made the ninety-foot fall, one on top of another; but the wind was whistling loudly, and even if they had heard it, it is unlikely that they would have understood its meaning. We have already observed that buffalo, especially when they are traveling in herds, are not given to reflection. To the very last one, they tumbled over; there was nothing to stop them. Later, the gigantic pile of skeletons—spread over an area of several hundred yards—was a sight which the inhabitants of the region flocked to see. There was not a single wolf's bone among them, and this fact enables us to reconstruct the end of the scene. The lean, gray hunters must have stopped at the edge of the cliff to sniff at the empty space, perhaps detecting, in spite of the snow, the odor of the fantastic banquet which awaited them, and seeking at once the paths that would lead them to it.

75

The Buffaloes of Africa

The forested plains of eastern and southern Africa are the home of the African Cape buffaloes (above and right). These ponderous animals grow to about five feet in height at the shoulder and weigh up to 2,000 pounds. Cape buffaloes are gregarious. They live in herds (right) that number anywhere from a dozen to hundreds of individuals, depending on the availability of food—primarily grasses and herbs—and water, on which they rely for bathing (above) as well as drinking. A typical herd, usually led by an older female, spends the early-morning hours drinking and foraging; the evening is given over to grazing. During the heat of the day the animals may seek the shade of the forest. They also loll in mud wallows to help lower body temperature and soothe skin that is irritated by blood-sucking ticks and insects. Buffaloes live together peaceably except during the breeding season, when bulls fight over females. Battles are usually ritualized, punctuated by threat displays and pushing contests. The dangerous tips of their horns are rarely used, and one combatant almost always retreats before blood is shed.

76

Exotic Bovines

The wild oxen shown on these pages demonstrate the extraordinary ability of their kind to adapt to extremes of climate. The Congo buffalo (below) is an African deep-forest dweller that has large ears with long white tufts that offset its resemblance to a domesticated cow. The water buffalo (opposite, below) is found from India to the Philippines, where it feeds on aquatic plants and grasses near rivers and lakes, lowland swamps and moist forests. To keep cool and thwart insects, water buffaloes use their horns to throw layers of mud over their backs.

The yak (opposite, above), which lives in the Himalayas at altitudes up to 20,000 feet, has a thick coat to protect it against the cold and broad hooves that enable it to roam nimbly about its rugged habitat.

Shaggy, mountain-dwelling yaks (right) and sleek, tropical water buffaloes (below) share heavy, crescent-shaped horns and similar size (bulls of both species may weigh close to a ton). As its name implies, the water buffalo is dependent on swamps, lakes and streams for wallowing and bathing as well as drinking. But the yak also needs and enjoys water. Even in the coldest weather it has been known to bathe in the icy lakes of its Tibetan habitat.

A small herd of water buffaloes fords a stream in Australia. These creatures, not indigenous to Australia, are descendants of domesticated Asian animals that were introduced into North Australia as farm animals in 1825.

Deer

From the majestic eight-foot-tall bull moose of Canada and Alaska to the 15-inch-high pudu of South America, most of the 53 species of deer are distinguished from all their relatives by their antlers. These splendid displays, which have made them the most sought-after game animals in many parts of the world, are usually grown by male deer. (Among the exceptions are reindeer and caribou—both sexes grow antlers—and the musk deer and Chinese water deer, which have tusks and no antlers.) Like the horns of antelopes, antlers serve both in male display and as weapons, but, unlike horns, antlers are shed annually and are regrown in time for the rutting season.

Antlers are unusual in aspects other than their beautiful branching structure, for nature has worked out a complex cycle of growth and regrowth. At the stage when they are most often used as weapons—in the competition for females during the rut—they are dead bone, which can be cut or broken off without pain or loss of blood. But when the antlers are "alive," they are growing, covered with tissue aptly called velvet that is pulsing with blood vessels and nerves and as sensitive as a drilled tooth.

The male deer lies low through the spring and summer months when this structure is developing and females are sexually unavailable. During that time the deer's blood is feeding bony material that may amount to one quarter of the animal's total skeletal weight to the buds that spring from the forehead of most species immediately after the previous season's antlers have been shed. Toward mid-summer the antlers of many species are full-grown but are still covered with the velvet. Within weeks this tissue begins to die and shred away from the underlying bone. The deer aids in the removal of the velvet by rubbing it off on the nearest shrub or tree. At this point the herbivorous animal may eat the dried velvet in order to replace the essential nutrients his body has lost during the highly demanding antler-growing process, which involves an enormous drain on the mineral resources of the buck's system as well as on his energy.

Despite all the trouble he goes to for his antlers, their value as a weapon for dominance during the rut and for protection against predators remains questionable. As a means of defense, antlers often appear to be inefficient. Among caribou and reindeer, for instance, winter is the most perilous time in the fight to survive against wolves, because the deep snow makes rapid flight impossible. Yet,

at this very time bulls shed their racks. On the other hand, among some European red deer the complete absence of antlers seems to offer no handicap—at least during the rut. Antlerless stags, called hummels, have had as much success in competing for females as their antlered brethren and have readily repelled rivals with sharp butting blows of their foreheads.

For the males of most deer species, the rutting season itself is a time of intense activity, stress and, in the end, exhaustion. The females have to put up with the harassment of bulls as they collect and herd their harems, but otherwise the season is relatively easy for cows.

In many species, the bulls announce the onset of the rutting season, often in late summer, by "roaring." The American elk, or wapiti, bugles, making a sound that is a combination bellow-whistle-grunt. This relative of the European red deer, the second largest deer on the continent after the moose, weighs up to half a ton, stands five feet tall at the shoulder and has antlers measuring up to five and one half feet across. Elk bulls will also flail around among bushes and small trees, smashing at the growth with their antlers. As they gather their harems, elk bulls eat little and expend enormous amounts of energy keeping the cows together and fending off challengers.

In these circumstances, perhaps it is just as well that during the rutting season individual elk cows are receptive to mating for only 17 hours at a stretch, with an interval of 21 days between estrous periods. Scent alerts the bull to a cow's receptivity. He approaches, stretches his neck and sets his chin down on the female's back. If she is indeed willing, mating takes place. A bull may mate with the same cow again, sometimes within the half hour. This repetition helps insure conception, as demonstrated by the high pregnancy rate among elk cows, which runs between 85 and 90 percent.

During the month of October, the wapiti rutting season tapers off, stopping completely toward the end of the month. Frazzled by the demands of policing their harems and checkmating the moves of brash bachelor bulls, the harem masters are thin, seedy shadows of their former arrogant selves. They hang their heads and move lethargically. Losing interest in the cows, they abandon their harems, which then may form the nuclei of cow herds, and join in winter herds with many of the same bulls they had so recently fought.

The Long Trek of the Caribou

During two major annual migrations, bands of barren ground caribou (right), the most numerous of four subspecies, start their trek across nothern Canada and Alaska to merge into herds that can range in size from less than 100 to thousands. Herds may cover 500 to 600 miles each spring and fall, with some backtracking of the spring migration trail in midsummer. Although the caribou's traditional routes may suddenly alter in response to some environmental change, their migration is not a haphazard search for food but a procession to a known destination along a definite trail established by generations of caribou.

Caribou disperse to feed, but at the slightest sign of danger they run together to form a compact band. It is only in an emergency that a caribou will gallop, reaching a maximum speed of more than 40 miles an hour—a pace that a caribou cannot sustain for long.

Within an hour after birth a newborn fawn can stand; two hours later it can run. A day-old calf can run faster than a man. In another two to three days the calf is ready to join the other animals (above). It will begin to graze when it is two to three weeks old, but most calves continue to make mother's milk their staple diet for three or four months after birth.

A Semiwild Servant

Although wild reindeer are still common in parts of the Eurasian tundra, for nearly a century most reindeer in the northern areas of Scandinavia inhabited by Lapps have been "domesticated." Actually, the Lapps' reindeer are still semiwild. Although the people tend their herds (left), they allow the reindeer to migrate over ancient routes, following the animals as the reindeer travel between summer and winter feeding grounds.

In this relationship, the Lapps feel a deep bond of dependence. As folklore would have it, the reindeer could get on well without them, but the Lapps could not survive without the reindeer. The reindeer supplies the Lapplanders with meat, dairy products, hide (for clothing and shoes), sinew (for thread) and a means of transportation. The reindeer is a fine draft animal—the only one among the species of deer. A single reindeer can carry 90 pounds in saddlebags, as much as 450 pounds on a sled over snow and, with suitable rest, can travel 40 miles a day.

Aptly named, reindeer lichen is a favorite reindeer food. Unfortunately, the lichens readily absorb radioactive fallout from atmospheric testing, notably strontium 90 and cesium 137. These radioactive isotopes are a potential threat to the reindeer-eating population of Scandinavia and the Soviet Union, who receive this radioactivity through the meat and dairy products they consume.

The Majestic Wapiti

Magnificent antlers with seven points (or tines) on each side signify that the wapiti stag shown opposite is in his prime. A wapiti stag will generally develop his first set of antlers about one year after birth. These are usually mere spikes. Each year after that, until he is seven to 12 years old, his new set of antlers will be progressively larger and grander until they have an average of six points on each side. As he grows older his antlers will normally diminish in size, and in time odd shapes may occur.

Wapiti cows suckle their newborn fawns (below) five or six times a day, and weaning does not take place until the fawn is three months of age or older. To allow mothers time off for their own essential feeding, wapiti cows have developed a cooperative baby-sitting system. While one or two mothers watch the young, the others are free to wander off and feed.

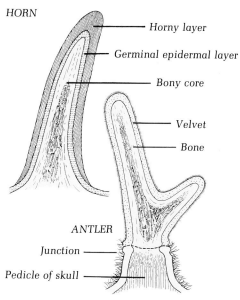

HORN

— Horny layer

— Germinal epidermal layer

— Bony core

— Velvet

— Bone

ANTLER

Junction —

Pedicle of skull —

Unlike horns (above), which are permanent appendages that grow throughout an animal's life and consist of a bony core covered by a tough outer layer, antlers are shed and regenerated annually and are composed of pure bone. While growing, antlers are covered with a velvety skin, which dries up and is rubbed off before the rut.

88

Spotted for Life

Most deer are spotted when they are born, but as they grow older their coats change to a solid white or chocolate brown, yellowish brown or grayish brown, often with distinct white markings. The axis deer (above), also called the chital, some fallow deer (left) and some subspecies of sika deer (opposite) are exceptions to this rule. They maintain their spots throughout their lives.

The axis deer, a native of India and Ceylon, is exceptional for another trait: the absence of a specific and limited rutting season. Though most other species of deer have an annual rut, usually lasting about two months, female chitals appear to be able to conceive during any month of the year, and newborn fawns may be seen all year long. Bucks with complete sets of hard antlers are present in the population throughout the year, a phenomenon that seems to coincide with a buck's readiness to mate.

There are 13 recognized subspecies of sika deer native to eastern Asia. The Japanese sika (right) is numerous in its home forests and parks, where it is the only indigenous deer. It has also proven adaptable wherever it has been introduced, including New Zealand, Europe and America. But Formosan sika deer, like the does shown below, are an insular race. Formerly abundant on Taiwan, they have become extinct due to excessive hunting for their meat and the bucks' antlers, which are said to be worth $100 a pair as trophies.

RED DEER

by Richard Jefferies

Richard Jefferies' humanitarian approach to the study of animals provides his readers with new insights into the beauty of a wild creature's everyday life. The description of a stag in the selection from Jefferies' book Red Deer *that follows is an almost mystical vision of the deer in its natural environment.*

There is no more beautiful creature than a stag in his pride of antler, his coat of ruddy gold, his grace of form and motion. He seems the natural owner of the ferny coombes, the oak woods, the broad slopes of heather. They belong to him, and he steps upon the sward in lordly mastership. The land is his, and the hills, the sweet streams and rocky glens. He is infinitely more natural than the cattle and sheep that have strayed into his domains. For some inexplicable reason, although they too are in reality natural, when he is present they look as if they had been put there and were kept there by artificial means. They do not, as painters say, shade in with the colours and shape of the landscape. He is as natural as an oak, or a fern, or a rock itself. He is earth-born—autochthon—and holds possession by descent. Utterly scorning control, the walls and hedges are nothing to him—he roams where he chooses, as fancy leads, and gathers the food that pleases him.

Pillaging the crops and claiming his dues from the orchards and gardens, he exercises his ancient feudal rights, indifferent to the laws of house-people. Disturb him in his wild strong-hold of oak wood or heather, and, as he yields to force, still he stops and looks back proudly. He is slain, but never conquered. He will not cross with the tame park deer; proud as a Spanish noble, he disdains the fallow deer, and breeds only with his own race. But it is chiefly because of his singular adaptation and fitness to the places where he is found that he obtains our sympathy.

The branching antlers accord so well with the deep shadowy boughs and the broad fronds of the brake; the golden red of his coat fits to the foxglove, the purple heather, and later on to the orange and red of the beech; his easy bounding motion springs from the elastic sward; his limbs climb the steep hill as if it were level; his speed covers the distances, and he goes from place to place as the wind. He not only lives in the wild, wild woods and moors—he grows out of them, as the oak grows from the ground. The noble stag in his pride of antler is lord and monarch of all the creatures left to us in English forests and on English hills.

RICHARD JEFFERIES (1848–1887)

In *Red Deer*.

Sheep and Goats

Separating the sheep from the goats is easy enough in the farmyard. But in the craggy mountainscapes where both animals live wild, they have assumed similar modes of existence and even closely resemble each other in appearance (though, unlike sheep, male goats usually have beards and tails that turn up). Both creatures, scientists believe, can be traced back to animals that lived some eight million years ago in Asia, members of a genus called *Oioceros*. The skulls of these ancient creatures had characteristics of both sheep and goats.

Goats and sheep, which share membership in the subfamily Caprinae within the family Bovidae, the horned ruminants, were domesticated about 10,000 years ago. While sheep seem less resourceful in fending for themselves as a result of their long association with man, goats can and do adapt more readily than sheep to life in the wild. Other members of the *Caprini* tribe are the tahrs (pages 100–101) and Barbary sheep (pages 104–105).

One goat, the ibex, was once believed to be useful in curing all sorts of maladies and misfortunes: Ibex blood was a palliative for calluses; rings made of their handsome 30-inch horns were thought to bestow good health; hair balls taken from ibex stomachs were deemed effective against cancer; its heart was a potent charm against bad luck. Because of their supposed supernatural attributes ibexes were hunted nearly to the point of extinction in the Swiss and Italian Alps and the Pyrenees of Spain, where they favor cool, windswept slopes above the tree line, feeding on sparse grass and herbs. (Nonhuman predators include wolves, bears and lynxes; all but the latter are scarce or absent from the ibex's habitat today.)

During the 19th century a rivalry over the remaining ibexes developed between Italy—where a remnant herd of 30 to 40 animals was placed under royal protection in the mid-1800s—and Switzerland, which tried unsuccessfully to obtain some of the offspring of the Italian herd. In 1906 some ibexes were secretly taken out of Italy into Switzerland. Today, as a result of breeding and releasing ibexes into the mountain wilds, the ibex population in the Alps is somewhere around 8,000, of which about 3,500 are found in Switzerland. But the Spanish ibex population has dwindled to fewer than two dozen creatures.

European ibex females and their young—usually born singly in May or June—remain in separate groups from the males, except during the winter breeding season. In December and January, bucks, which grow as tall as three feet and weigh up to 240 pounds, begin to battle one another for females, rising on their hind legs and crashing their horns together.

Biologists consider the tahr a near kin to the goat, calling it a goat-antelope. Its name, *Hemitragus*, means semigoat, and indeed the Himalayan tahr looks like a goat wearing some other creature's red-brown fur coat. It grows about a yard high and has no beard. Tahrs usually cluster in small groups, but members of the family in the Far East often gather into herds of 30 or 40.

Like the tahr, the Rocky Mountain wild goat is considered a goat-antelope. The Rocky Mountain goat, *Oreamnos americanus*, can grow as large as 40 inches in height at the shoulder and weigh as much as 200 to 300 pounds. It can be found from Cook Inlet, Alaska, to Idaho. Both sexes are horned and bearded. They live in small herds, though males sometimes are loners. Mountain goats extend the surefootedness of domestic goats, which have been seen to walk along the chrome side trim of parked automobiles, a mere fraction of an inch of horizontal surface. Their hooves more often allow them to find footholds on apparently sheer cliff faces where such predators as wolves cannot follow. Probably the North American mountain goat's worst adversary is an avalanche, roaring down the mountain with a velocity even the goat cannot escape.

Bighorn sheep, which live on the escarpments of North American mountain ranges and in Siberia, form groups that can number as many as 50. Bighorns are equipped with skulls made of two layers of connecting bone. The construction acts like a shock-absorbing helmet, enabling the bighorn rams to engage with impunity in stupendous bouts of head-butting, impacting in head-on charges that can be heard a mile away, all in order to establish dominance and the right to mate with ewes.

Today the bighorn is an endangered animal—some subspecies are extinct—because of its inability to coexist with man and his livestock. Bighorn sheep are confined to a rugged, rocky habitat, acquiring their knowledge of their home terrain from the flock and lacking the capacity to pick up knowledge of new terrain by themselves. This peculiarity stymies wildlife management efforts to increase bighorn populations. In frustration, one authority (Wayne King of New York's Bronx Zoo) has even suggested that "transplanted animals should literally be led around and acquainted with every piece of useful habitat in the region they are to occupy."

Rocky Mountain goat

Intrepid Alpinists

The Rocky Mountain goat, the most agile of all large North American mammals, spends its summers at elevations of 10,000 feet or more. It enjoys the cool air and the available forage of the Rockies and of mountain ranges from Alaska to Idaho. Life among the rugged cliffs often above the timberline has the advantage of being relatively predator-free, since bears, wolves and cougars with designs on the mountain goat simply can't match its astonishing climbing skills.

As with most animals that live largely untroubled by predators, however, internecine battles are correspondingly more severe. The goats' nine-inch horns are sharp and can be deftly wielded to puncture an opponent. Grizzlies have been known to die in struggles with mountain goats. When two males square off to contest for a female, the results may be fatal, and neither antagonist is likely to emerge unscathed.

Grazing in Olympic National Park in Washington, mountain goats take their nourishment from grass and other low-lying vegetation. When winter snows arrive, the animals are often forced to lower altitudes to find food. There they are even more vulnerable to what naturalists believe are the greatest natural killers of mountain goats: avalanches and rock slides, caused by thaws, heavy rains or triggered by the animals themselves.

97

High above a cloud-filled gorge in Washington State (left), mountain goats occupy a tiny alpine meadow. Below, a mountain goat rests on a promontory overlooking a precipitous drop. Mountain goats don't scramble over the rocks when alarmed or pursued, but move with the confident gait of accomplished mountaineers, trotting boldly beside thousand-foot drops on ledges only inches wide. Such daring is possible largely because of the mountain goat's concave hoof, which has a tough rim surrounding a soft, cushioned pad that grips rock surfaces.

Ibexes (above) graze in Gran Paradisio National Park in the Alps of northern Italy. Ibexes range from western Europe across the Near East and into the Caucasus, with two subspecies surviving in Africa. Those in the European Alps live above the timberline, at altitudes of 12,000 feet or higher. A Cretan wild goat (left), a dainty and colorful wild goat subspecies, pauses in its meanderings. Inhabitants of the Greek island since the Minoan age, they suffered severe losses when firearms were introduced. Today, they live in a national park established for their protection, and they have been declared the national animal of Greece.

Rare Goat Relatives

The animals on these pages, members of the same subfamily as goats and sheep, share their relatives' preference for mountains. A creature of Asian and Middle Eastern highlands, the tahr (right) resembles a true goat in both habit and appearance but is equally related to sheep and goats. Men of the hills where they live hunt them for food and regard their meat as a remedy for rheumatism and fever. Tahrs travel in herds of 30 to 40 animals and are extremely wary creatures. At least one or more individuals are on the lookout for danger when they stop to feed—on almost any plant life they can find—or to rest.

The chamois (below), a graceful mountain inhabitant of Europe, is distinguished by slender horns that are set close together and bend backward at the top to form hooks. Its skin, once used for cleaning and buffing metal, is now made into luxury clothing.

Bighorns' Weighty Weaponry

The magnificent horns of the North American bighorn sheep pictured above are both the instruments of its survival and of its demise. The curled horns, which weigh from 18 to 30 pounds, serve mature males as weapons and symbols of dominance among the socially gregarious bighorns. But they are also highly coveted by hunters. A poacher's black market in bighorn trophies has only recently been substantially curtailed.

Bighorn sheep range from British Columbia south into Mexico, but populations are sparse. Expanded cattle and sheep ranching in the American West during the early part of this century resulted in overgrazing and proved perilous to bighorns, which were driven out of their habitats.

During the summer, bighorn rams live near ewes and their young but do not join them. In the fall mating season, rams square off in horn-clashing duels to establish dominance and breeding rights. The losers in these contests are so dominated that they are even mounted occasionally by higher-ranking males.

Predators—including man—find it no easy work to pursue the bighorn. With its superior hearing, smell and eyesight (said to be acute up to five miles) the bighorn generally avoids confrontations with possible predators. If it must face an adversary, the sheep will exercise one of two options—fight or run. Few animals can withstand the freight-train rush of a charging bighorn, and fewer still can pursue it as it speeds over rocks, jumping as far as 17 feet over chasms a thousand feet deep.

Two rams settle their differences (left) in a crunching battle of horns. The contest begins with both animals running forward to pick up speed. Then they rear up on their hind legs. Finally, falling to all four, they meet with a deafening crash. Special double-layered skulls and a tough exterior hide prevent serious injury. Below, the victor (left) of one such duel chases the loser through a band of ewes.

Three Survivors

Rough, inhospitable tracts of land are the habitats of many species of wild sheep. These sheep gather in small groups, sometimes consisting of only an adult male and female and their offspring, perhaps because such barren surroundings cannot support large aggregations of animals. One such animal is the Barbary sheep (left), or aoudad, whose shawl of long hair, extending from its neck to its forelegs, distinguishes it from all other sheep. It is found in the arid areas of northern Africa, where, to compensate for lack of water, it is able to survive on the dew and moisture extracted from the green vegetation it feeds on.

The mouflon (right), smallest of the wild sheep, prefers the dry, precipitous ridges of the mountains of Sardinia and Corsica. It is the only wild sheep that, in winter, grows a wooly undercoat, similar to the familiar fleece of domestic sheep. The long curved horns of the Dall sheep (below) are much like those of the bighorn (pages 102–103) but are smaller and wider set. Found in northwest Canada and Alaska, Dall sheep can also be distinguished from bighorn sheep by their pure white coat.

Pronghorns and Other Survivors

There are a number of herd animals that still roam free today only because they were saved from extinction by the efforts of man—after decades of uncontrolled hunting had brought them close to the vanishing point. Among them are the pronghorn of the American West, the musk-ox of the far Arctic north and the saiga of the Russian steppes.

Pronghorns were indiscriminately slaughtered for food by pioneers moving westward, by farmers and ranchers protecting their range against the competitive grazing of animals, and by sportsmen. As a result, herds that once numbered in the millions counted less than 20,000 by 1908. Since that time, efforts at conservation, including strict hunting controls, have restored the numbers of pronghorns in the western United States, Canada and Mexico to nearly 400,000.

The first whites venturing into the West incorrectly identified pronghorns as antelopes. They are not true antelopes but are instead the only members of the family Antilocapridae; they differ from antelopes in shedding their horns. Pronghorns stand about a yard high, males weighing between 100 and 140 pounds, does about 10 to 20 pounds less. Their main protection is superb eyesight—likened to 8x binoculars; white erectile hairs on their rumps, which they can flash like a heliograph to signal danger; and their great speed. Pronghorns are the fastest land mammals on the North American continent and the swiftest in the world after the cheetah. Pronghorns can maintain a pace of 40 miles per hour for as much as five miles at a stretch and can run at 30 miles per hour over longer distances.

Pronghorns live in herds whose composition and numbers shift with the seasons of the year. Large groups gather during the winter, then fragment into small aggregations of females and their young, while males form bachelor herds. As the breeding season of late summer approaches, males begin assembling harems of up to 15 does.

The musk-ox was named by an 18th-century French explorer who asserted—incorrectly—that the animal secreted musk, an ingredient in perfume then much in demand (and supplied by musk deer and civets). In claiming another source for the aromatic substance, the explorer hoped to attract financial backing for his expeditions.

The animal he named is a massive, shaggy creature that stands four to six feet high at the shoulder and can weigh up to 900 pounds. It originated somewhere in north-central Asia more than a million years ago and dispersed westward into France and England and eastward into Siberia, crossing to North America via the Bering Strait land bridge about 90,000 years ago. The animals were nearly extirpated by modern Eskimo hunters and commercial fur traders.

The musk-ox has an outer coat of coarse hair and an undercoat of soft, silken wool, which insulates the animal against the cold, enabling it to survive in its forbidding habitat. (Arctic nights can last almost five months, and winter temperatures range from zero to $-70°$ F.)

For protection against wolves, musk-oxen cluster in a circle formation, calves in the center. Under attack, they dart out of the circle at their enemies, slashing with front hooves, then retreat to protect the calves. Since musk-oxen do not run away, this defensive tactic worked well not only against wolves but also men with lances. Men with guns, however, could pick the animals off one by one.

Hunters killed the musk-oxen in such numbers that in 1917 the Canadian government began a program to save the remaining few thousand animals. Today some 10,000 musk-oxen live in preserves in northern Canada. Another 6,000 live in Greenland in herds of more than 100 animals. Attempts are being made to domesticate musk-oxen to facilitate the harvesting of the fine woolly undercoat.

The saiga antelope, an ungulate that grows only 29 inches tall at the shoulder and weighs a mere 95 pounds, was saved from possible extinction when, in 1919, the Russian government prohibited its hunting. Before that, demand for the saiga's foot-long accordion-ribbed horns, widely believed to be of medicinal value, had severely diminished the numbers of this animal, once widespread over the entire area from Poland to the Caucasus Mountains. Spotty enforcement of the hunting ban during the first chaotic years of the Soviet Union further reduced the population so that only a few hundred animals survived by 1930. In the next two decades, better enforcement was a major factor in restoring the population to some three million head.

But perhaps the most important cause of the saiga's comeback was its fertility. Female saigas can mate at the age of seven months, before their skeletal development is complete; and unlike most other hoofed animals, which usually have only one offspring at a time, they deliver twins in 65 percent of their births.

Pronghorn antelopes

Big-hearted and Playful

The white hairs of its rump patch erect, signaling a potential threat, a pronghorn trots across an Arizona range (opposite, below). These animals are physically well adapted for speed, with a large windpipe and lungs for the intake and absorption of oxygen and a large heart that quickly pumps oxygenated blood throughout the body. A buck pronghorn's heart is twice as large as that of a sheep of equal body weight. Pads of cartilage on their front feet help pronghorns travel across wild, rocky terrain.

Even when not alarmed, pronghorns are likely to spurt suddenly across the range for no reason other than sheer exuberance; they seem to enjoy racing one another. Herds develop intricate games of follow-the-leader that are enjoyed by adults as well as young. Pronghorns have been known to race an automobile or a horse, a contest that usually ends when the pronghorn abruptly crosses in front of the car or horseman. But because the pronghorns are so nimble and fast, they are rarely struck by cars.

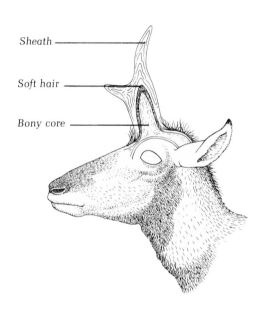

Sheath

Soft hair

Bony core

The pronghorn is named for the characteristic hooked horns borne by both males and females of the species (right), although the doe's horns are generally slimmer and much shorter than the buck's and seldom protrude further than the length of its ears. The unique construction of a pronghorn's semipermanent horn is shown in the cutaway drawing above. Like true antelopes, the pronghorn maintains a solid bony core throughout the year. Building on this foundation, the animal grows a sheath of hairlike tissue that develops into horny branches which the pronghorn, unlike the antelope, sheds annually.

Crossing a Montana range, two does and a buck (at right in the picture above) pause to scan the route ahead. When a group of pronghorns is on the move a doe (right) generally takes the lead, and a buck brings up the rear to prod laggards. Thus, even from a distance, the positions of does and bucks help a viewer to distinguish males from females.

Back from the Brink

Despite three decades of protection, in 1954 musk-oxen were still threatened by extinction. In an effort to rescue them from this plight, scientists began experiments to tame these unusual animals. To secure calves for breeding, researchers, with the permission of the Canadian government, hunted them by helicopter, swooping down and scattering the herds like the one shown above, separating the youngsters from the protection of the herd. Then, men on foot grappled with the calves and captured them.

Today musk-oxen are being selectively bred in Alaska, Norway and Canada. Domestication is not without its problems. To prevent the musk-oxen bulls from goring one another, their horns (left) must be removed. Even then the animals' heads must be insulated with old tires; otherwise the bulls would injure themselves in fights. But the musk-oxen possess a resource worth all this trouble—a silky, cashmerelike undercoat of hairs called *qiviut* that can be harvested from the live animal.

Standing in a defensive posture, a captive bull and cow flank their calf. With the increased quality of nutrition they get in captivity, musk-ox cows breed annually. In the wild, when nourishment is scarce, reproduction may be restricted to alternate years.

The Feral Herds

Among the most problematic creatures on earth are the so-called feral herds, descendants of domesticated animals which either escaped or were released into the wild and now live in an untamed state. These feral herd animals include goats, cattle, burros, water buffaloes, horses, sheep, pigs and camels.

The routes these animals traveled to find their way back to the wild from their domestic state are numerous and diverse; in many cases, they overlap. Basically, however, creatures can become feral in three ways:

●Introduction or importation into regions where they have not previously been found in nature, as meat on the hoof or as beasts of burden. In the 18th and 19th centuries, seafarers introduced goats to the island of Catalina, off the California coast, and to the Hawaiian chain. The purpose of this importation was to have meat available in mid-voyage without the necessity of carrying it aboard ship. Left untended, the goats swiftly multiplied. In Australia and New Zealand, settlers imported water buffaloes from Asia to help plow fields and to drag heavy loads. Some of these buffaloes wandered away from the settlements and formed the nuclei of feral herds.

●Dispersion over wide, unfenced ranges. The mustangs of the American West, descendants of horses imported primarily by the Spanish in the 16th century or escapees from ranches and farms, were allowed to graze over vast open stretches of prairie, and consequently many were lost to their human masters. Their descendants form the core of bands that still roam the shrinking open space on the far western plains. Burros, also introduced to North America by the Spanish, became feral in much the same manner.

●Maintenance in an untamed state as a pool of available animals for meat or for use as draft or saddle stock. In this category fall the horses and cattle of the Camargue, a wild, marshy island in southern France, and the ponies named for the island of Chincoteague off the Virginia coast, where they appeared in Colonial times, perhaps as refugees from a wrecked ship. Today, they live on the nearby island of Assateague, a relatively wild environment to which the animals were transferred in the 19th century when Chincoteague became built up.

With these animals, the line between the domestic and feral states is ambiguous. Camargue horses (pages 114–115) were once widely used as cart pullers in France, and the cattle were often incorporated in domestic herds. Chincoteague ponies have been traditionally rounded up each year and many broken for riding.

The cattle of the Camargue and the feral burros that live in California's Death Valley demonstrate behavioral patterns suggesting that domestication goes no deeper than man-made contrivances—fences and tethers. A Swiss ethologist named R. Schloeth spent some 2,000 hours between 1956 and 1961 studying the Camargue cattle and concluded that their behavior matched that of truly wild cattle in highly significant respects. He learned, for example, that the Camargue cattle, like wild oxen, live by their own societal rules, evinced by myriad assertions of dominance and corresponding indications of subservience. As in the wild, top-ranking individuals hold their heads high when confronting inferiors, which drop their heads and lick the dominant animals submissively on the shoulder.

Observers of Death Valley burros have noted comparable analogies between their behavior and that of wild equines. Each male burro, or jack, establishes dominance over a territory (usually a quarter square mile in extent), allowing other males to cross the area to drink but withdrawing the privilege and challenging encroaching males when a jenny or a female ready for mating is present.

To some observers, feral animals evoke a mystique of liberty triumphantly regained. To others, however, they represent economic and ecological disruption, for example, as horses and burros compete with range cattle and sheep for forage.

Even the mustang, that graceful, haunting reminder of America's past, has been indicted as a pest by cattle and sheep men and some conservationists as well, who condemn the animal for grazing off the scarce range fodder. As a result, mustangs have been shot, driven over cliffs and slaughtered for pet food by the thousands. Laws to protect mustangs from this sort of treatment have been in force since 1971. Ironically, they have been so effective that the mustang population, currently about 60,000, is deemed too large even by the standards of the protectionists. The law does allow destruction of the animals in case of overpopulation after all other remedies (including adoption procedures) have been exhausted—and though many may deplore it, this last-ditch solution may have to be invoked.

112

A mustang family

Lords of the Camargue

Inhabiting one of Europe's most spectacular nature preserves on a delta in the south of France, herds of Camargue horses spend most of their lives galloping through the delta's uncultivated marshlands (right) and freely grazing on its rough pastures. They never see the inside of a stable. Yet despite this apparent freedom, their lives are subtly monitored. These beautiful creatures (above and on the cover) straddle the murky dividing line between wild and domesticated animals.

In a truly wild state, natural selection would determine the mating of stallions and mares. In the Camargue, local ranchers intervene. Each year they round up three- and four-year-old stallions and from among them choose those they wish to breed. The chosen ones are turned loose to rejoin the mares. The others are trained for herding, most of them after castration.

Each day after work the horses are smacked on the rump and turned back to feed in the wild. Next morning, when the herdsman wants his horse, he must go to the marsh to separate him once again from his feral companions and catch him.

114

The Mustangs

by J. Frank Dobie

As an amateur naturalist, J. Frank Dobie was primarily interested in the animals of the southwestern United States and studied them in detail. In this selection from his book The Mustangs, *Dobie describes Starface, the leader of a band of mustangs—a horse with a will and a strength that both challenged and terrified the local ranchmen.*

Starface was a deep bay with a white star-shaped patch in his forehead and a stocking on his right forefoot. It was believed that he had Morgan blood in his veins; he might have been pure Spanish. In 1878 he was commanding a large band of mustangs that ranged between the Cimarron and Currumpa rivers in No Man's Land—the westward-pointing panhandle of Oklahoma. Every step that Starface took was a gesture of power and pride.

He would have been a marked horse anywhere, but his character made him even more noted than his carriage. He was the boldest gallant and the most magnificent thief that the Cimarron ranges had ever known. Most ranchmen in No Man's Land had horse herds as well as cattle and some raised horses altogether. Whenever Starface felt the blood stir in him, he would raid down upon these ranch horses, fight off the domestic stallions, cut out a bunch of mares with or without colts, and herd them back into his own well-trained bunch. The country was as yet unfenced, and Starface knew it all, claimed it all. He became a terror.

No man could walk him down, for Starface refused to circle. Nor could any man get near enough to crease him. Finally, the harassed ranchers organized to capture him. They took hundreds of long-distance shots at him; they cut off most of his followers; but he still ran free. Then they picked four cowboys, furnished them with the strongest and fastest horses in the country, and told them not to come back until they had killed or captured Starface. The four scouted for nearly a week before they sighted Starface's band. By keeping out of sight and riding in relays, they dogged the suspicious mustangs for three days and nights.

Most of the time the cowboys kept back in the edge of the breaks on the south side of the Cimarron.

They studied the habits of mustangs as they had never studied them before. They marveled at the discipline by which the stallion kept his band in order. Now he would leave them and graze off alone, and not a mare would dare follow. Now he would round them into a knot that no yearling dared break from. Again he would course out with every animal obediently at his heels. Starface seemed to require less sleep than any other horse of the band.

It was early fall and the moon was in full quarter. Shortly after midnight on the fourth night the two cowboys on watch saw Starface leave his mares and head for the river flats. One man followed while his partner sped back to arouse their companions. A light dew on the grass made trailing easy; besides, the stallion was so intent on his quest that he seemed to pay no regard to what might be behind him. For six miles he galloped into the north.

Then, about ten miles east of the present town of Kenton, he entered a grassy canyon. Spreading out between walls of rock on either side, this canyon narrows into a chute that, in time of rains, pitches its waters off a bluff into the Cimarron River. Not far above the brink at the canyon mouth Starface passed through a narrow gateway of boulders shutting in a small valley.

Daylight was not far away when the cowboys came to the pass. They were familiar with the boxed structure of the canyon below them. They knew that Starface would before long be returning with his stolen mares. They decided to wait for him. They were sure that their opportunity had come. They were all determined to catch Starface rather than kill him, for studying him had changed vengeance into admiration.

In the early light they watched the bold stallion maneuvering about a dozen mares and colts. They were untrained to his methods, and Starface was wheeling and running in every direction, checking his captives at one point and whipping them up at another. Like a true master, he was intent on his business—and, for once, he was off guard.

He had worked the bunch into the pass, where the walls were hardly a hundred feet apart, and now the mares were stringing into discipline, when suddenly the four cowboys dashed from behind the boulders. Pistol shots shook the morning stillness. The wild Texas yell frenzied even the dullest of the mares. Ropes slapped against leather leggins and sang in the air.

With a wild snort of challenge, Starface charged alone up the steep canyon side. At first the cowboys thought he had discovered a trail out unknown to them. They stood still, watching, not a gun drawn. As the mustang ascended into a patch of sunshine allowed by a break in the walls on the opposite side of the canyon and they could see the sheen of light on his muscles, one of them called out, "God, look at the King of the horse world!" Long afterwards in describing the scene he added, "Not a man at that moment would have shot that animal for all the horses north of Red River."

But only for a brief time were they doubtful of capturing the superb stallion. They saw him leap to a bench as wide perhaps as a big corral—wide enough for a reckless cowboy to rope and manage an outlaw mustang upon. Towering above that bench was the caprock, without a seam or a slope in its face. Starface had picked the only spot at which the bench could be gained. But, like the canyon floor he had fled from, it ended in space—a sheer jump of ninety feet to the boulder-strewn bed of the Cimarron.

"Come on, we've got him," yelled one of the mustangers.

Under the excitement, the horses they were riding leaped up the way the mustang had led. Now he was racing back and forth along the bench. As the leading rider emerged to the level, he saw Starface make his last dash.

He was headed for the open end of the bench. At the brink he gathered his feet as if to vault the Cimarron itself, and then, without halting a second, he sprang into space. For a flash of time, without tumbling, he remained stretched out, terror in his streaming mane and tail, the madness of ultimate defiance in his eyes. With him it was truly "Give me Liberty or give me Death."

The Decline of a Legend

In 1900 there were an estimated two million untamed mustangs roaming the sparsely settled vastness of the American West, in part descendants of tough Spanish horses imported by the conquistadores in the 16th century. For 200 years the mustangs increased in the wild; later they were caught and tamed in great numbers by Indians and colonists. Mustang stock supplied the Pony Express, the United States Cavalry and many a cowpoke.

The four horses galloping through the brush above are among the 50,000 mustangs currently inhabiting nine western states. Their numbers were sharply reduced after World War II, when they were systematically hunted to supply meat for dog-food canneries. By 1970 their population had dwindled to an estimated 17,000. In 1971 a federal law was passed to protect the mustangs, and according to many ranchers, the law has done its job all too well. They claim that the present mustangs are inbred—worthless, untrainable descendants of runaways and abandoned domestic animals with little trace of the original Spanish bloodline. But horse lovers, who helped pass the legislation, maintain that a mustang's worth cannot be measured by words or dollars.

Object of bitter controversy for many years among local ranchers, land conservationists, hunters and wild-horse lovers, a band of mustangs (right) runs through the rugged Pryor Mountains of northern Wyoming, now a wild-horse refuge. According to a special report commissioned by the Department of Interior, the mustang is the creature best suited to this area's poor habitat. In other parts of the West, ranchers are still sharply critical of the proliferation of mustangs on public lands, where the ranchers, who pay for the right to graze livestock, consider the mustangs trespassers.

While mustang mares graze placidly, seemingly unaware that they are the cause of the commotion, two stallions (left) rear up on their hind legs and begin to pound at each other with their hooves. Stallions often try to avoid such a confrontation by ritual posturing, eyeball to eyeball. But when this fails, the combat that ensues is often a bloody one—even a fight to the finish.

Credits

Cover—T. Nebbia, DPI. 1—W. Ruth, Bruce Coleman, Inc. 5, 6–7—George Calef. 9—T. McHugh, Photo Researchers, Inc. 12–13—S. Wilson, Entheos. 15—John Dominis, Time Inc. 16—R. Kinne, P.R., Inc. 17—M. Levy, P.R., Inc. 18, 19—C. Bavignoli, Time Inc. 20 (top)—S. Trevor, B.C., Inc., (bottom, left)—H. Albrecht, B.C., Inc., (bottom, right)—S. Trevor, B.C., Inc. 21—(top, left)—R. Kinne, P.R., Inc., (top, right)—G. Shapira, P.R., Inc., (bottom, left)—S. Trevor, B.C., Inc., (bottom, right)—Leonard Lee Rue III, B.C., Inc. 22–23 (top)—T. Nebbia, DPI. 22—John Dominis, Time Inc. 23—John Dominis, Time Inc. 24–25—John Dominis, Time Inc. 30–31—Leonard Lee Rue III, P.R., Inc. 31—John Dominis, Time Inc. 32—M.N. Boulton, Nat'l. Audubon Society Collection, P.R., Inc. 32–33—Tatarsky, DPI. 35—N. Myers, B.C., Inc. 37—E. Appel, P.R., Inc. 38—C. Haagner, B.C., Inc. 38–39—N. Myers, B.C., Inc. 40 (left)—Leonard Lee Rue III, B.C., Inc., (right)—K.W. Fink, B.C., Inc. 41 (top)—G.D. Plage, B.C., Inc., (bottom)—Leonard Lee Rue III, B.C., Inc. 42 (top)—N. Myers, B.C., Inc., (bottom)—R.D. Estes, P.R., Inc. 43 (top)—M. Quraishy, B.C., Inc., (bottom)—N. Myers, B.C., Inc. 44—Peter B. Kaplan. 45 (top)—Nina Leen, (bottom)—C. Bavagnoli, Time Inc. 47—E. Appel, P.R., Inc. 48 (top)—G. Holton, P.R., Inc., (bottom)—Bruce Coleman, Inc. 49 (left)—Bruce Coleman, Inc., (right)—D. Steffen, P.R., Inc. 50–51—T. Hollyman, P.R., Inc. 52–53—M.P. Kahl, B.C., Inc. 53 (top)—Bruce Coleman, Inc., (bottom)—F. Erize, B.C., Inc. 54–55—Des & Jen Bartlett, B.C., Inc. 56—F. Erize, B.C., Inc. 56–57—R. Peterson, P.R., Inc. 59—H. Reinhard, B.C., Inc. 60—H. Reinhard, B.C., Inc. 61 (top)—J. Foott B.C., Inc., (bottom)—N. Myers, B.C., Inc. 63—J. Van Wormer, B.C., Inc. 64—Bruce Coleman, Inc. 64–65—J.M. Burnley, B.C., Inc. 76—E.R. Degginger, B.C., Inc. 77—B. Campbell, B.C., Inc. 78—Peter B. Kaplan, P.R., Inc. 79 (top)—M.N. Boulton, B.C., Inc., (bottom)—K. Tanaka, Animals Animals. 80–81—T. Nebbia, DPI. 83—Leonard Lee Rue III, B.C., Inc. 84—S. Wilson, DPI. 84–85—N. Devore III, B.C., Inc. 86–87—J. Verde, P.R., Inc. 87—K. Brate, P.R., Inc. 88—H. Engels, Nat'l. Audubon Society Collection, P.R., Inc. 89—J.M. Burnley, Nat'l. Audubon Society Collection, P.R., Inc. 90 (top)—M.P. Kahl, B.C., Inc., (bottom)—H. Reinhard, B.C., Inc. 91 (top)—J. Burton, B.C., Inc., (bottom)—R. Kinne, P.R., Inc. 93—J. Hancock, P.R., Inc. 95—B. Brooks, B.C., Inc. 96 (top)—John Dominis, Time Inc., (bottom)—Dain, P.R., Inc. 97—K. Gunnar, B.C., Inc. 98–99—K. Gunnar, B.C., Inc. 99—J. Foott, B.C., Inc. 100 (top)—P. Thatcher, P.R., Inc., (bottom)—K. Fink, P.R., Inc. 101 (top)—K. Fink, B.C., Inc., (bottom)—E. Duscher, B.C., Inc. 102—D. Hiser, P.R., Inc. 103 (top)—Bob & Clara Calhoun, B.C., Inc., (bottom)—W. Fraser, B.C., Inc. 104—Bruce Coleman, Inc. 105 (top)—T. Branch, P.R., Inc., (bottom)—W. Ruth, B.C., Inc. 107—Leonard Lee Rue III, B.C., Inc. 108—K. Fink, Nat'l. Audubon Society Collection, P.R., Inc. 109 (top)—N. Mishler, P.R., Inc., (bottom)—J. Couffer, B.C., Inc. 110–111—L.B. Jennings, Nat'l. Audubon Society Collection, P.R., Inc. 110—K. Fink, Nat'l. Audubon Society Collection, P.R., Inc. 111—F. Erize, B.C., Inc. 113—Wolfgang Bayer. 114—T. Nebbia, DPI. 114–115—H. Sylvester, Rapho, P.R., Inc. 120–121—B. Eppridge, Time Inc. 121—N. Devore III, B.C., Inc. 122–125—Wolfgang Bayer. 128—W. Ruth, B.C., Inc.

Photographs on endpapers used courtesy Time-Life Picture Agency, Russ Kinne and Stephen Dalton, P.R., Inc., and Nina Leen.

Film sequence p. 8—"Mustang," a program in the Time-Life Television series Wild, Wild World of Animals.

ILLUSTRATION p. 10—L. Raboni; p. 11—C. Tarka; pp. 26–29—C. Robinson; pp. 66–68—The Bettmann Archive; pp. 72–75 and 117–119—John Groth; pp. 88 and 108—Enid Kotschnig. Painting pp. 70–71—Nat'l. Collection of Fine Arts, Smithsonian Institution.

Bibliography

NOTE: Asterisk at left means that a paperback volume is also listed in Books in Print.

Allen, Thomas B., The Marvels of Animal Behavior. National Geographic Society, 1972.

Bannikov, A. G., "The Last Refuge of the Wild Camel." International Wildlife, January 1976, p. 31.

———, Zhirov, L.V., Lebedeva, L.S., and Fandeev, A. A., Biology of the Saiga. Israel Program for Scientific Translation, 1967.

Bere, Rennie, Antelopes. Arco, 1970.

Burton, Maurice, Systematic Dictionary of Mammals of the World. Thomas Y. Crowell, 1962.

Cahalane, Victor H., ed., Alive in the Wild. Prentice-Hall, 1969.

Caras, Roger A., North American Mammals. Meredith Press, 1967.

Clark, James L., The Great Arc of the Wild Sheep. University of Oklahoma Press, 1964.

Coblentz, Bruce, "Wild Goats of Santa Catalina." Natural History, June/July 1976, p. 70.

Cott, Hugh B., Looking at Animals. Charles Scribner's Sons, 1975.

*Darling, F. Fraser, A Herd of Red Deer. Doubleday, 1964.

*Dary, David A., The Buffalo Book. The Swallow Press, 1974.

D'Aulaire, E., and D'Aulaire, O., "Still a Dream Machine." International Wildlife, January 1975, p. 29.

Dobie, Frank, The Mustangs. Clarkson Potter, 1952.

Einarsen, Arthur S., The Pronghorn Antelope. Wildlife Management Institute, 1948.

Ensminger, M. E., Animal Science. Interstate, 1969.

Franklin, William L., "High, Wild World of the Vicuna." National Geographic, January 1973, p. 77.

Geist, Valerius, The Mountain Sheep. University of Chicago Press, 1971.

Groves, Colin P., Horses, Asses and Zebras in the Wild. Ralph Curtis Books, 1974.

Grzimek, Bernhard, Grzimek's Animal Life Encyclopedia, Vols. 12, 13. Van Nostrand Reinhold, 1975.

*Haines, Francis, The Buffalo. Thomas Y. Crowell, 1970.

Hyams, Ed, Animals in the Service of Man. J. B. Lippincott, 1972.

Kramer, Raymond J., Hawaiian Land Mammals. Charles E. Tuttle, 1971.

Leakey, Louis S. B., Animals of East Africa. National Geographic Society, 1969.

Leonard, Arthur Glyn, The Camel. Longmans, Green, 1894.

McCullough, Dale R., The Tule Elk: Its History, Behavior and Ecology. University of California Press, 1971.

McHugh, Tom, The Time of the Buffalo. Alfred A. Knopf, 1972.

Mochi, Ugo, and Carter, T. Donald, Hoofed Mammals of the World. Charles Scribner's Sons, 1971.

Moehlman, Patricia des Roses, "Getting to Know the Wild Burros of Death Valley." National Geographic, April 1972, p. 502.

Mohr, Erna, The Asiatic Wild Horse. J. A. Allen, 1971.

Morris, Desmond, The Mammals. Harper & Row, 1965.

Moss, Cynthia, *Portraits in the Wild*. Houghton Mifflin, 1975.

Murie, J. Olaus, *The Elk of North America*. The Stakepole Company, 1951.

Park, Ed, *The World of the Bison*. J. B. Lippincott, 1969.

Reader's Digest Association, *The Living World of Animals*. Reader's Digest, 1970.

Rearden, Jim, "Caribou, Hardy Nomads of the North." *National Geographic*, December 1974, p. 858.

Rorabacher, J. Albert, *The America Buffalo in Transition*. North Star Press, 1971.

Rue, Leonard Lee, *The World of the White-tailed Deer*. J. B. Lippincott, 1969.

Ryden, Hope, *America's Last Wild Horses*. E. P. Dutton, 1970.

Simpson, George Gaylord, *Horses*. Oxford University Press, 1951.

Talbot, L. M., and Talbot, M. H., *The Wildebeest in Western Masailand*. Wildlife Monograph, 1963.

Teal, John J., "Domesticating the Wild and Woolly Musk Ox." *National Geographic*, June 1970, p. 862.

Van Wormer, Joe, *The World of the American Elk*. J. B. Lippincott, 1969.

———, *The World of the Pronghorn*. J. B. Lippincott, 1969.

Vaughan, Terry A., *Mammalogy*. W. B. Saunders, 1972.

Whitehead, G. Kenneth, *Deer of the World*. Viking, 1972.

Index